A DISSECTION GUIDE AND ATLAS TO THE FETAL PIG

David G. Smith
University of Oklahoma

Michael P. Schenk
University of Mississippi Medical Center

Morton Publishing Company
925 W. Kenyon Avenue, Unit 12
Englewood, Colorado 80110

10 9 8 7 6 5 4 3 2 1

Printed in the United States of America

ISBN: 0-89582-363-2

PREFACE

This dissection guide is intended to provide a comprehensive introduction to the anatomy of the fetal pig. This manual bridges the gap between the traditional atlas format, utilizing photographs and illustrations, and the laboratory manual format, which relies heavily on descriptive text, by integrating the two styles into one convenient source. Its concise design allows it to be used as a supplement to other laboratory manuals that may be used in your course. Its depth is sufficient to thoroughly cover all the major organs and organ systems at a level consistent with the curriculum of most beginning biology and zoology laboratories at the university or advanced high school level. The order of coverage of the different systems follows a logical progression that maximizes the ease with which students can dissect relevant structures.

The text is informative, highlighting material that can be applied to other life science courses. Chapters begin with lists of objectives to focus student's attention on the "big picture" of each chapter. Full-color photographs and illustrations provide accurate representations of the anatomy to facilitate identification of anatomical structures. Frequent comparisons between the fetal pig and humans are given to compare and contrast the anatomy and physiological processes of the fetal pig and those in humans (e.g., fetal vs. adult circulation). Color micrographs of histological sections of relevant tissues accompany many of the photos and illustrations to give students an appreciation for the microanatomy associated with different tissues and organs, further emphasizing the comparative approach employed throughout the book. Tables are used to conveniently summarize information presented in the text. Dissection instructions are *italicized* and set off from the text, while important terms are **boldfaced**. A glossary containing definitions of boldfaced terms is provided, along with a detailed index for quick reference.

ACKNOWLEDGMENTS

Without the efforts of many dedicated individuals, this manual, which has burgeoned into a beautifully crafted book of color photos, illustrations, tables and text, would have remained an impossible idea. In sincere appreciation of these efforts, the authors would like to thank Chris Rogers and others at Morton Publishing for the opportunity to put this guide together as well as Joanne Saliger for working with us on the page layouts and formats. We would like to pay special thanks to Darryl C. Smith, M.D., for his valuable editorial comments and constructive criticisms that strengthened both the accuracy of the content and the organization of the text, and to Bill Armstrong for his excellent, detailed photography of the dissected pig.

CONTENTS

Introduction 1

BASIC DISSECTION TECHNIQUES 2
BODY PLANES AND REGIONS 2

1 External Anatomy...5

LABORATORY OBJECTIVES.....................................5
GENERAL EXTERNAL FEATURES.................................5
FEMALE EXTERNAL FEATURES..................................6
MALE EXTERNAL FEATURES....................................6

2 Fetal Skeleton 9

LABORATORY OBJECTIVES.....................................9
AXIAL SKELETON...11
APPENDICULAR SKELETON....................................11
TYPES OF JOINTS..12

3 Muscular System...13

LABORATORY OBJECTIVES....................................13
THE NECK...14
    Superficial Musculature................................14
    Deep Musculature......................................17
PECTORAL REGION AND FORELIMB............................18
    Superficial Musculature................................18
    Deep Musculature......................................21
THE ABDOMEN...22
THE PELVIC REGION AND HINDLIMB.........................23
    Superficial Musculature................................23
    Deep Musculature......................................25

4 Digestive System 31

LABORATORY OBJECTIVES....................................31
HEAD, NECK AND ORAL CAVITY.............................31
ABDOMINAL CAVITY..34

5 **Cardiovascular System**..................................**43**

LABORATORY OBJECTIVES...43
THORACIC CAVITY AND NECK REGION 43
   The Heart 43
   Veins of the Thoracic Region 48
   Arteries of the Thoracic Region 49
ABDOMINAL CAVITY 53
   Hepatic Portal System..53
   Arteries and Veins of the Abdominal Region 58
   Umbilical Cord 60

6 **Respiratory System**....................................**65**

LABORATORY OBJECTIVES...65
THE THORACIC CAVITY..65
THE ORAL CAVITY..67

7 **Reproductive and Excretory Systems**.........**71**

LABORATORY OBJECTIVES...71
MALE REPRODUCTIVE SYSTEM 72
FEMALE REPRODUCTIVE SYSTEM 75
EXCRETORY SYSTEM...78

8 **Nervous System** **83**

LABORATORY OBJECTIVES...83
THE BRAIN...83
THE EYE 87

9 **Endocrine System** **91**

LABORATORY OBJECTIVES...91
NECK REGION 92
ABDOMINAL REGION..92

References **95**

Glossary **97**

Index **103**

INTRODUCTION

This dissection guide is intended to provide an introduction to the anatomy of the fetal pig for beginning students in biology and zoology laboratories at the university or advanced high school level. The fetal pig is an excellent organism for the study of vertebrate anatomy due to its similarities to humans and other mammals, its manageable size, its low cost and its availability. These fetuses come from mature female sows that, when slaughtered for their meat, are discovered to contain unborn young; they are not raised specifically for dissection purposes. Therefore their use as dissection specimens does not promote unnecessary killing of animals for scientific purposes. The gestation period of these animals is typically 114 days (16 weeks). The pigs you will dissect are probably between 100 and 114 days old (8–12 inches in length) and possess fully-developed organs and organ systems.

While the fetal pig is often used in laboratory classes as a comparison to humans, it should be noted that there are differences in the anatomy and physiological processes of the fetal pig and those in humans. As a result, we have given some examples of comparisons between fetal pigs and humans to avoid confusion over these cases (e.g., fetal vs. adult circulation). Nonetheless, because of our common ancestry, pigs and humans do possess a majority of characteristics in common, and many homologous structures are shared between pigs and humans.

Many features of this dissection guide allow the student quick access to the information presented and should facilitate use of this manual.

1. Each chapter begins with a list of objectives.
2. Color photographs and illustrations are provided for identification of anatomical structures.
3. Color micrographs of histological sections of relevant tissues accompany many of the photos and illustrations.
4. Tables are used throughout to conveniently summarize information presented in the text.
5. Dissection instructions are *italicized* and designated with an arrow, while important terms that students should remember are **boldfaced.**
6. A glossary containing definitions of boldfaced terms is provided for quick reference.

Since for many students this will be their first major dissection, we will review proper dissection techniques and the terminology associated with the orientation of body planes and regions.

Basic Dissection Techniques

Practice safe hygiene when dissecting. Wear appropriate protective clothing, gloves and eyewear, and DO NOT place your hands near your mouth while handling preserved specimens. While many of the preservatives currently used are non-toxic to the skin, they may cause minor skin irritations in some individuals and definitely should not be ingested.

Read all *italicized* instructions CAREFULLY before making any incisions. Make sure you understand the direction and depth of the cuts to be made — many important structures may be damaged by careless or imprecise cutting. For instance, while investigating the digestive system, you do not want to damage structures in the circulatory system that you will need to see later.

Use scissors and your dissecting probe whenever possible, unless otherwise instructed. Despite their popularity, scalpels usually do more harm than good and should not be relied upon as your primary dissection tool.

When instructed to "expose" or "view" an organ, you should attempt to remove all of the membranous tissues that typically cover these organs (fat, fascia, etc.). Your goal should be to expose the organ or structure as completely as possible. Many arteries and veins are embedded deeply in other tissues, while muscles are grouped closely together. These structures will require careful "cleaning" to adequately identify them.

A good strategy to use when working in pairs is to have one student read aloud the directions from the book while the other student performs the incisions. These roles may be traded from section to section to give both students a chance to participate. In addition to simply identifying the organs and structures from the photos/illustrations, make sure you read the descriptions of them in the text. You should be able to recognize each organ *and* describe the function it performs in the body.

Body Planes and Regions

The following terms will be used to refer to the regions of the body and the orientation of the organs and structures you will identify. A section perpendicular to the long axis of the body separating the animal into cranial and caudal portions is called a **transverse** plane. The terms **cranial** and **caudal** refer to the head and tail regions, respectively. A longitudinal section separating the animal into right and left sides is called a **sagittal** plane. The sagittal plane running down the midline of the animal has a special name, the median plane. Structures that are closer to the median plane are referred to as **medial**. Structures further from the median plane are referred to as **lateral**. **Dorsal** refers to the side of the body nearest the backbone, while **ventral** refers to the side of the body nearest the belly. A longitudinal section dividing the animal into dorsal and ventral parts is called a **frontal** plane. **Proximal** refers to a point of reference nearer to the dorso-ventral midline of the body than another structure (e.g., when your arm is extended, your elbow is proximal to your hand). **Distal** refers to a point of reference farther from the midline of the body than another structure (e.g., when your arm is extended, your elbow is distal to your shoulder). **Rostral** refers to a point closer to the tip of the nose.

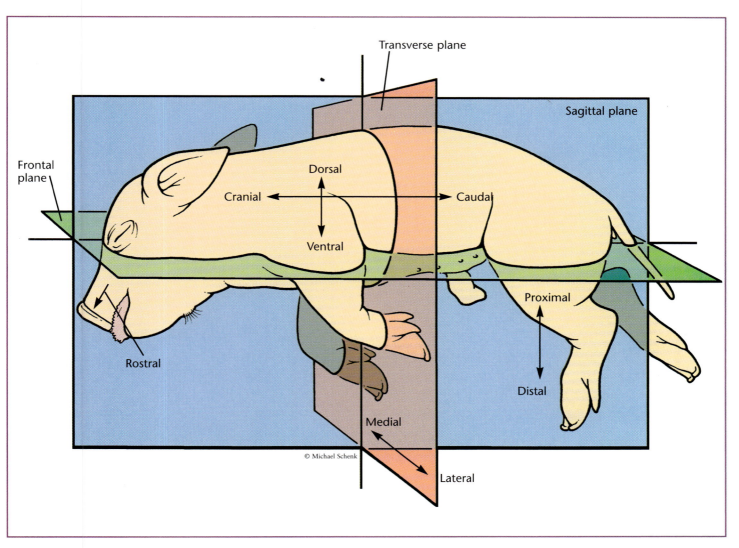

Transverse plane

Sagittal plane

Frontal plane

Dorsal

Cranial ←——————————————→ Caudal

Ventral

Rostral

Proximal

Distal

Medial

Lateral

© Michael Schenk

🔷 **INTRODUCTION**   *Illustration of anatomical planes of reference and regions of the body on a quadrupedal animal.*

CHAPTER 1

EXTERNAL ANATOMY

Laboratory Objectives

After completing this chapter, you should be able to:

1. Identify the major external features of the fetal pig.
2. Identify the function of all indicated structures.
3. Determine the sex of your pig and the external structures unique to males and females.

General External Features

> *Obtain a fetal pig from your instructor. Position your pig on its side in a dissecting pan so that you may observe the external features of your pig.*

The body of the pig is divided into the **head** and **trunk** regions (Figure 1.1 and 3.2). The trunk is further divided into the **thorax** and **abdomen** regions (which are separated internally by the diaphragm). The thorax houses the heart and lungs, while the abdominal region houses the major digestive, excretory and reproductive organs of the pig. Notice the sensory organs concentrated around the head. There are **eyes**, which are not yet open, **ears**, **nares** (for sensing chemicals dissolved in the air) and **vibrissae** (commonly called whiskers, for tactile sensations). These organs all play important roles in the pig's ability to sense and respond to stimuli in its environment.

> *Turn your pig over on its dorsal side, so that you may view the structures on its ventral surface.*

Identify the **umbilical cord** protruding from the ventral side of the abdomen. This structure carries nutrient- and oxygen-rich blood to the fetus and removes excess metabolic waste products and carbon dioxide from the fetal system. You will learn about the internal structures of the umbilical cord in Chapter 5. You should be able to determine the sex of your pig using external features. While

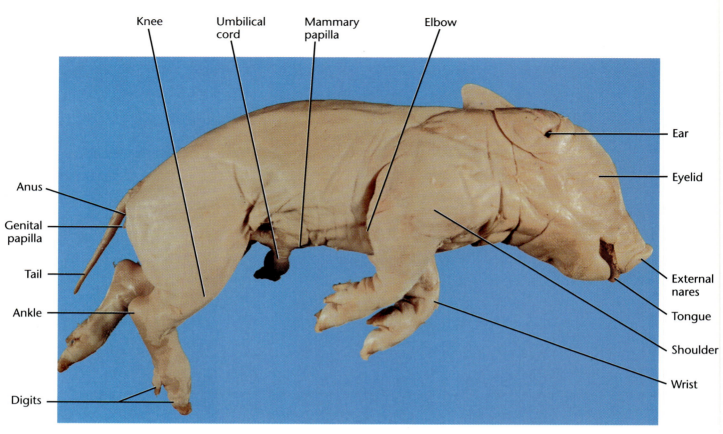

Knee   Umbilical   Mammary   Elbow
cord   papilla

Ear

Eyelid

Anus

Genital
papilla

External
nares

Tail

Tongue

Ankle

Shoulder

Wrist

Digits

🔺 **FIGURE 1.1**   *Lateral view of the pig showing external features.*

both sexes have many of the same structures, some of their locations differ in males and females. Both sexes have **mammary papillae** around the abdominal region near the umbilical cord (Figure 1.2). In females, these will develop into the mammary glands and will be used to provide milk for their newborn young. While males do possess these structures, they do not provide any known function. Both males and females possess an **anus**, located just ventral to the base of the tail. It is through the anus that undigested foodstuffs are eliminated from the body.

Female External Features

Females have a **urogenital opening** ventral to the anus, near the base of the tail (Figure 1.2a). This represents the opening to the reproductive pathway and serves as a channel for the release of excretory products (urine) from the body. This opening may be slightly obscured by the **genital papilla**.

Male External Features

In males, the **urogenital opening** is located on the ventral surface just caudal to the umbilical cord (Figure 1.2b). This is the opening of the urethra which releases excretory products (urine) and semen in the adult pig. At this point, the penis is not fully developed but lies underneath the tissues of the abdomen along the ventral surface. If your male pig is old enough, there may be **scrotal sacs** located near the anus. As development proceeds, the testes, which originally

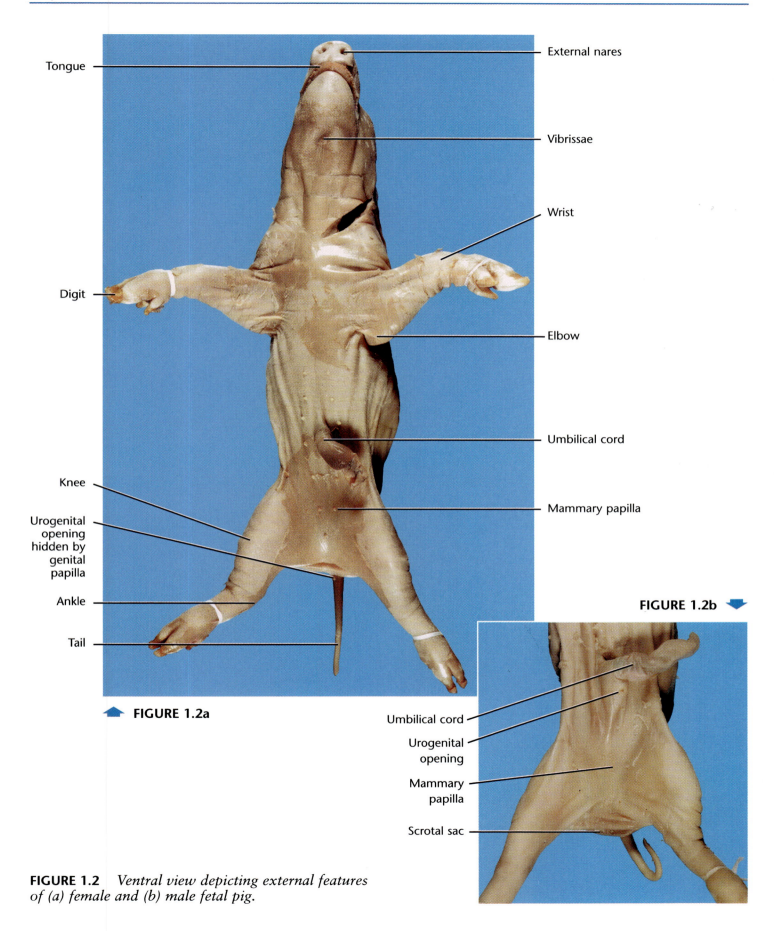

Tongue

External nares

Vibrissae

Wrist

Digit

Elbow

Umbilical cord

Knee

Mammary papilla

Urogenital
opening
hidden by
genital
papilla

Ankle

Tail

FIGURE 1.2a

FIGURE 1.2b

Umbilical cord

Urogenital
opening

Mammary
papilla

Scrotal sac

FIGURE 1.2   *Ventral view depicting external features
of (a) female and (b) male fetal pig.*

form deep inside the abdominal cavity near the kidneys, migrate caudally and eventually descend into the scrotal sacs. Since sperm production is highly sensitive to temperature, the testes of most mammals are housed outside of the body where temperatures are cooler than in the abdominal cavity. In humans, the temperature inside the testes is about 2°C cooler than the temperature inside the abdominal cavity. This is a precarious situation though, since environmental temperatures may drop too low. In this case, a special set of muscles known as the cremaster muscles retracts the testes, pulling them closer to the body to conserve heat. In many mammals, the testes only descend during breeding seasons.

FETAL SKELETON

Laboratory Objectives

After completing this chapter, you should be able to:

1. Identify the major skeletal features of the fetal pig.
2. Identify the different types of joints and discuss the movements they allow.

The skeletal system of vertebrates plays an important role in supporting the body and holding animals upright, yet it must allow for flexibility so that animals can perform a wide array of motions. Thus, while the skeletal system is composed of many individual calcified bones that are quite rigid, there are many different kinds of joints connecting these bones which permit movement. The skeletal system of chordates is called an **endoskeleton** since it is buried inside the animal, covered by layers of muscle and skin. One benefit that this type of skeleton affords is the ability for animals to grow without having to "shed" their old skeleton (as many arthropods with exoskeletons must do).

Since the skeletal elements of mammals are only beginning to form in the fetal stage, your pig will not have a fully-developed skeleton. At this point, many of the "bones" are not yet ossified but exist as the cartilaginous precursors to bones (Figure 2.1). As the pig matures, most of these sections of cartilage will slowly be replaced with bone as it grows outward from the center of each bone toward the joints. Therefore, you will not dissect your pig to study the skeletal system. If available, you may use a mounted skeleton of another mammal (e.g., cat or rat) to provide an example of a completely ossified skeleton. Compare this skeleton to the illustration of the fetal skeleton depicted in Figure 2.2. Since all mammals share a common ancestry, you will be able to see many homologous bones in the different mammals' skeletons. **Homologous structures** are structures in different species that are similar because the animals share a common ancestry. This principle forms the basis for the field of comparative anatomy — a branch of zoology that uncovers the evolutionary relationships among related groups of animals by studying their anatomical similarities and differences.

The skeleton is comprised of two different regions: the axial skeleton and the appendicular skeleton. The **axial skeleton** consists of the skull, vertebral

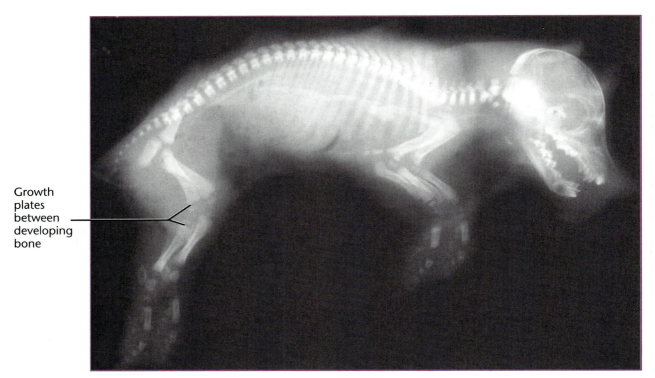

Growth plates between developing bone

▲ **FIGURE 2.1**   *X-ray of fetal pig showing the degree of skeletal development just prior to birth.*

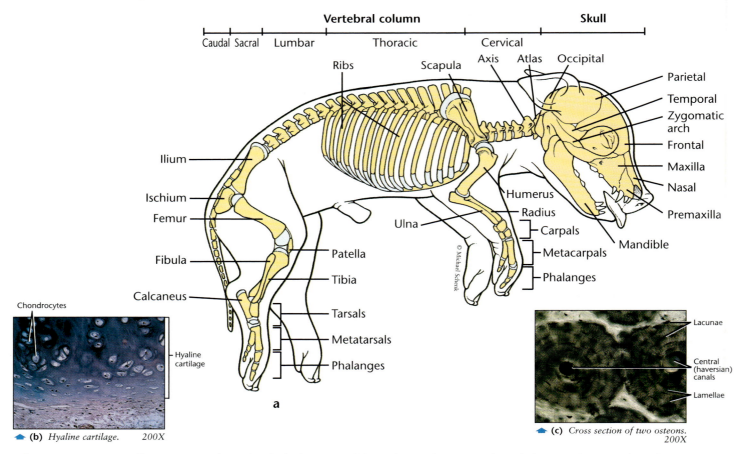

Vertebral column

Skull

Caudal   Sacral   Lumbar   Thoracic   Cervical

Ribs   Scapula   Axis   Atlas   Occipital

Parietal

Temporal

Zygomatic arch

Frontal

Maxilla

Nasal

Premaxilla

Mandible

Ilium

Ischium

Femur

Fibula

Calcaneus

Humerus

Radius

Ulna

Carpals

Metacarpals

Phalanges

Patella

Tibia

Tarsals

Metatarsals

Phalanges

© Michael Schenk

a

Chondrocytes

Hyaline cartilage

▲ **(b)** *Hyaline cartilage.*   *200X*

Lacunae

Central (haversian) canals

Lamellae

▲ **(c)** *Cross section of two osteons.*
200X

▲ **FIGURE 2.2**   *Illustration of (a) fetal skeleton and histology photographs of (b) cartilage and (c) bone.*

column and the rib cage. It forms the longitudinal axis of the body. The **appendicular skeleton** consists of the bones of the forelimbs and hindlimbs as well as the bones that attach them to the axial skeleton, the pectoral and pelvic girdles.

Axial Skeleton

The **skull** is actually comprised of several bones held together by immovable sutures along the surfaces of the bones. As such, it forms a rigid, protective covering for the delicate brain and sense organs of the pig. The lower jaw in mammals is composed of a single bone called the **mandible**. This is one characteristic that separates mammals from all other classes of vertebrates (like reptiles and birds). The most rostral bone in the upper jaw region is the **premaxilla**, which supports the incisors and canines. The **maxilla** supports the premolars and molars. Cranial to these two bones is the **nasal** bone which covers the snout region. The actual brain case is composed of four major bones: the **frontal** bone, **parietal** bone, **occipital** bone and **temporal** bone. The lateral sides of the skull are supported by the **zygomatic arch** which extends caudally from the orbit toward the base of the skull.

The skull joins the vertebral column at the first cervical vertebra, called the **atlas**. This is a highly specialized bone designed to fit precisely into the convex bulges in the base of the skull known as the **occipital condyles**. This type of joint allows a full range of motion in the head. The second cervical vertebra is the **axis**. In all, pigs have 7 cervical vertebrae. The thoracic region is composed of 14–15 **thoracic vertebrae.** Notice the many **ribs** extending from these vertebrae and enclosing the chest region. These ribs provide protection and support for the heart and delicate lungs which lie inside the thoracic cavity. Caudal to the thoracic vertebrae are the 6–7 **lumbar vertebrae**. These vertebrae have no ribs extending from them. A special group of 4 vertebrae are fused together to form the **sacrum,** an especially strong region that supports the pelvic girdle and hindlimbs. Finally, the **caudal vertebrae** continue from the base of the sacrum to the tip of the tail. In pigs, there are anywhere from 20–23 caudal vertebrae.

Appendicular Skeleton

Protruding from the thoracic region are the bones of the forelimbs. The **scapula**, or shoulder blade, forms the base of the forelimb. This bone is not actually attached to the axial skeleton; rather it floats in the muscle layers surrounding this region. This allows quadrupedal mammals to have a very fluid running motion (although if you've ever seen a pig run, you would probably not characterize its motion as fluid!). The proximal portion of the forelimb is comprised of the **humerus**. The distal portion of the forelimb is comprised of two bones, the **radius** and the **ulna**. The bones of the wrist region are known as the **carpals,** while the **metacarpals** and **phalanges** make up the forefoot.

The two major bones of the pelvic girdle in the pig are the **ilium** and **ischium**. The **femur** is the proximal hindlimb bone, while the **tibia** and **fibula** are the more distal hindlimb bones. Notice that the knee region has a small bone covering the juncture of the femur and the tibia and fibula. This is the **patella** or "knee cap." The "ankle" bones of the hindlimb are called the **tarsals,** and the bones of the foot are the **metatarsals** and **phalanges**. The large bone in the hindfoot that forms the slight bulge in the back of the hindlimb in mammals is the **calcaneus** bone. This is homologous to our heel bone. Pigs have a form of locomotion known as **digitigrade** locomotion, which means they literally walk on their digits — the tips of their "fingers" and "toes." Humans display **plantigrade** locomotion, meaning we walk on the soles of our feet (our weight is supported primarily by our metatarsals and tarsals, rather than our phalanges).

Types of Joints

There are several different ways in which bones join together to form articulations. The type of joint present reflects both the kinds of movement that the particular joint will permit and the amount of strength the joint provides for support. The different classes of joints found in the pig are summarized in Table 2.1.

TABLE 2.1 *Types of joints found in the pig and examples of where they occur in the body.*

Joint	Description	Example
SUTURE	Immovable connections between bones with interlocking projections; provides highest degree of strength but allows no motion.	Cranial surfaces
HINGE	Convex surface of one bone fits into concave surface of another; permits movement in only one plane.	Metacarpal/Phalange
SPHEROIDAL (Ball-and-socket)	Round head fits into cup-shaped socket; permits greatest range of motion.	Humerus/Scapula Femur/Ischium
GLIDING	Flat or slightly curved surfaces oppose one another for sliding motion; permits only slight movement, but in all directions.	Between Carpals Between Tarsals
PIVOT	One bone turns around another bone as its pivot point; permits rotating movements.	Radius/Ulna Atlas/Axis
CONDYLAR	Two knuckle-shaped surfaces engage corresponding concave surfaces; permits movement in only one plane.	Femur/Tibia

CHAPTER 3

MUSCULAR SYSTEM

Laboratory Objectives

After completing this chapter, you should be able to:

1. Identify the major muscles of the fetal pig.

2. Identify the actions of selected muscles in the fetal pig.

3. Discuss the different types of movements that muscles perform.

Muscles are designed with one basic purpose in mind — movement. Muscles work to either move an animal through its environment or move substances through an animal. In vertebrates, there are three basic types of muscle tissue — **skeletal muscle**, called striated muscle, **cardiac muscle** and **smooth muscle**, sometimes called visceral muscle (Figure 3.1). Some of these muscles, like skeletal muscle, can be voluntarily controlled by the animal, while others, like cardiac and many smooth muscles, produce actions that are involuntary. The muscles that you will dissect will be the skeletal muscles associated with the axial and appendicular portions of the skeleton. The musculature of vertebrates is quite complex and requires patience and care to properly dissect each muscle away from its nearby structures. It is often difficult to tell where one muscle ends and another begins, which is compounded by the fact that many muscles occur in groups. You should pay careful attention to the direction the muscle fibers run. Often this will give you clues as to where two muscles cross or abut. Another aspect to note is the origin and insertion of each muscle. The **origin** is the less movable location on a bone where a muscle attaches, while the **insertion** is typically the more movable attachment. Sometimes muscles attach to tendons instead of attaching directly to bone. The direction in which a muscle exerts force also plays a role in its shape, where it inserts and originates (and sometimes its name). A muscle that **adducts** moves a limb toward the midline of the body. Conversely, a muscle that **abducts** moves a limb away from the midline of the body.

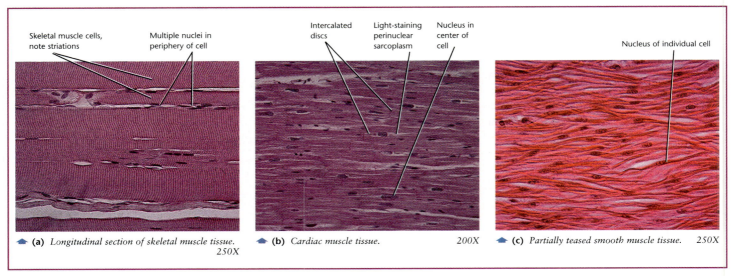

Skeletal muscle cells, note striations

Multiple nuclei in periphery of cell

Intercalated discs

Light-staining perinuclear sarcoplasm

Nucleus in center of cell

Nucleus of individual cell

(a) *Longitudinal section of skeletal muscle tissue.* 250X

(b) *Cardiac muscle tissue.*   200X

(c) *Partially teased smooth muscle tissue.*   250X

FIGURE 3.1   *Histology photographs of the three types of muscle tissue: (a) skeletal, (b) cardiac, and (c) smooth.*

Before you begin this section, you must first remove all of the skin from your pig. This process will take some time. The best strategy for removing the skin is to make the incisions indicated in Figure 3.2 and then use a blunt probe to tease the skin away from the underlying muscles. After the skin is removed, you may need to clean away the thin, membranous fascia that is still covering the muscles. Lay your pig on its side to view the superficial lateral muscles.

The Neck

Superficial Musculature

Examine the muscles of the pig's neck (Figures 3.3, 3.4, and 3.5). The largest and most obvious muscle is the **masseter**, the primary muscle involved in chewing. Caudal and dorsal to the masseter, you will find the **temporalis** and **sternomastoid**, which both act on the flexing of the head. Ventral to the masseter (on the underside of the neck), locate the **mylohyoid**. This is a long, thin muscle which runs longitudinally along the underside of the neck. The **digastric** muscle is located along the medial side of the mylohyoid. The **sternohyoid** is another long, thin muscle that runs longitudinally along the ventral side of the neck but is caudal to the mylohyoid. The **omohyoid** is a larger muscle that lies underneath the mylohyoid. The **brachiocephalicus** complex is actually a group of muscles (the **cleidomastoid** and **cleidooccipitalis**) which assist in pulling the forelimbs of the pig toward the head during activities such as walking and digging. The brachiocephalicus is located behind the ear, running from the base of the skull to the upper front shoulder. The **supraspinatus** is a fairly small muscle located at the point where the brachiocephalicus and the deltoid overlap. The supraspinatus also assists in extending the shoulder away from the body.

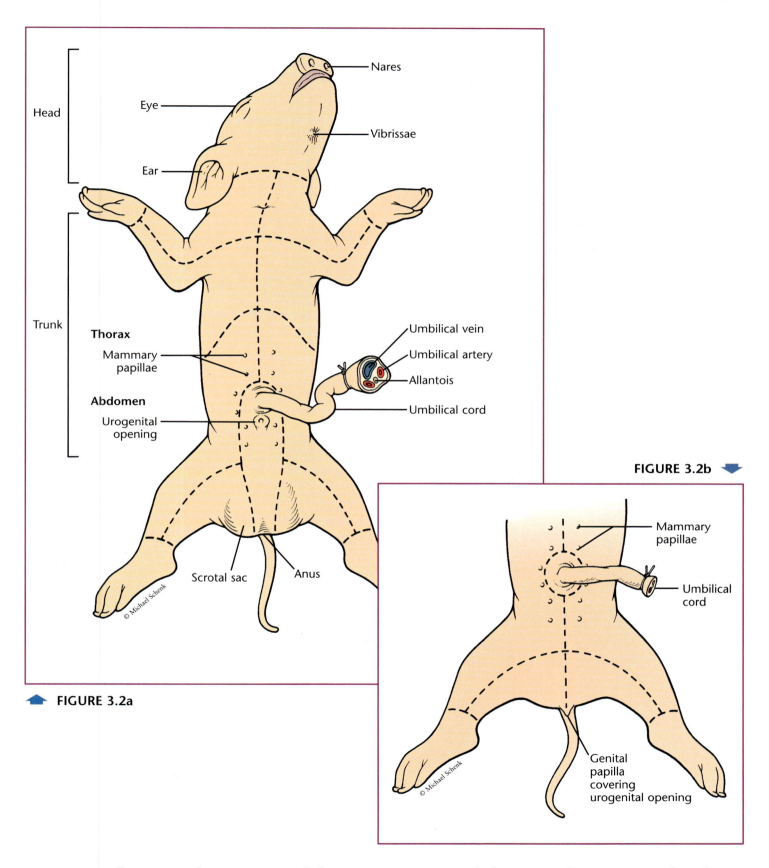

FIGURE 3.2b ➡

FIGURE 3.2a ⬅

FIGURE 3.2   *Illustrations showing suggested skin incisions to view underlying musculature on (a) male and (b) female.*

TABLE 3.1 *Superficial muscles of the neck. Refer to Figures 3.3, 3.4, and 3.5.*

Muscle Name	Action
Masseter	Elevates mandible.
Temporalis	Flexes head.
Sternomastoid	Flexes head.
Mylohyoid	Raises floor of mouth.
Digastric	Depresses mandible.
Sternohyoid	Pulls hyoid bone and tongue caudally.
Omohyoid	Draws hyoid bone caudally; retracts roof of tongue.
Brachiocephalicus: (Cleidomastoid and cleidooccipitalis)	Moves forelimb cranially.
Supraspinatus	Extends shoulder.

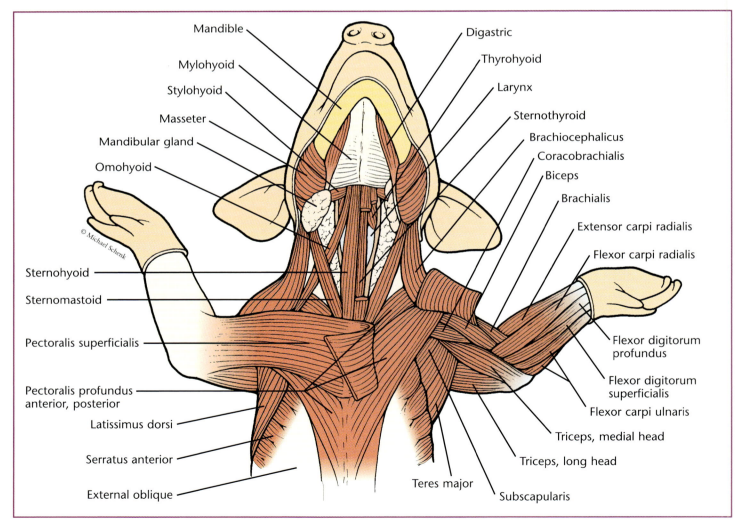

FIGURE 3.3 *Illustration of ventral view of neck depicting (pig's right) superficial and (pig's left) selected deep musculature of this region.*

Deep Musculature

On the ventral side of the neck use scissors (or a scalpel) to cut through the sternohyoid and omohyoid muscles to expose the underlying musculature.

Identify the **sternothyroid** muscle lying directly underneath the sternohyoid that was cut (Figure 3.3). This muscle moves the larynx caudally during vocalizations. Underneath the cut omohyoid, you should be able to locate the small **thyrohyoid** muscle which moves the hyoid bone caudally and dorsally.

On the lateral side of the head, locate the **temporalis** muscle. This muscle lies dorsal to the masseter and somewhat beneath the eye socket and ear area. The temporalis pulls the mandible cranially as the pig chews. Lying underneath the large masseter muscle (and protruding from the ventral border) is the **stylohyoid** muscle, another small muscle which controls the movements of the hyoid bone and tongue.

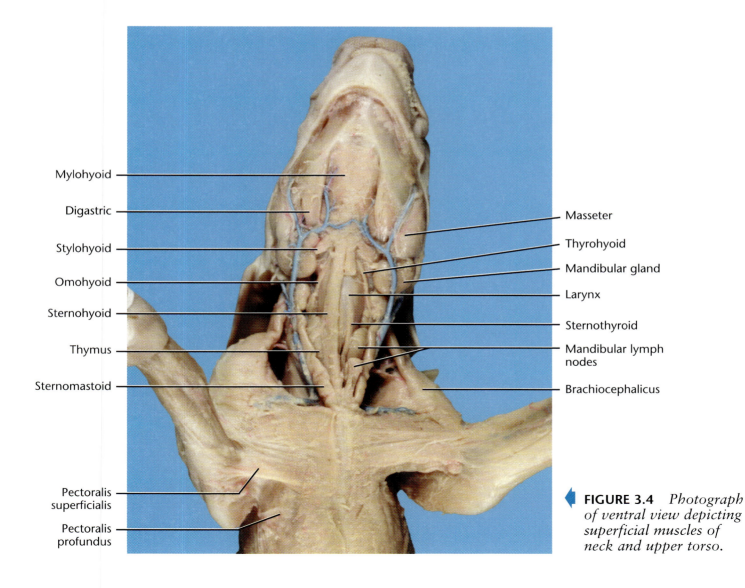

Mylohyoid

Digastric

Stylohyoid

Omohyoid

Sternohyoid

Thymus

Sternomastoid

Pectoralis superficialis

Pectoralis profundus

Masseter

Thyrohyoid

Mandibular gland

Larynx

Sternothyroid

Mandibular lymph nodes

Brachiocephalicus

FIGURE 3.4 *Photograph of ventral view depicting superficial muscles of neck and upper torso.*

TABLE 3.2　*Deep muscles of the neck. Refer to Figures 3.3 and 3.5.*

Muscle Name	Action
Sternothyroid	Moves larynx caudally.
Thyrohyoid	Moves hyoid caudally and dorsally.
Temporalis	Pulls the mandible upward.
Stylohyoid	Raises (lifts) the hyoid; draws base of tongue and larynx dorsally and caudally.

Pectoral Region and Forelimb

Superficial Musculature

On the lateral side of the thoracic region, locate the **pectoralis profundus** (Figures 3.5–3.7). This muscle passes along the crest of the scapula and attaches to the supraspinatus at its cranial end. The other end attaches to the humerus and adducts the forelimb of the pig. Just caudal to the pectoralis profundus, locate the **omotransversarius**, a narrow muscle that extends between the atlas and the

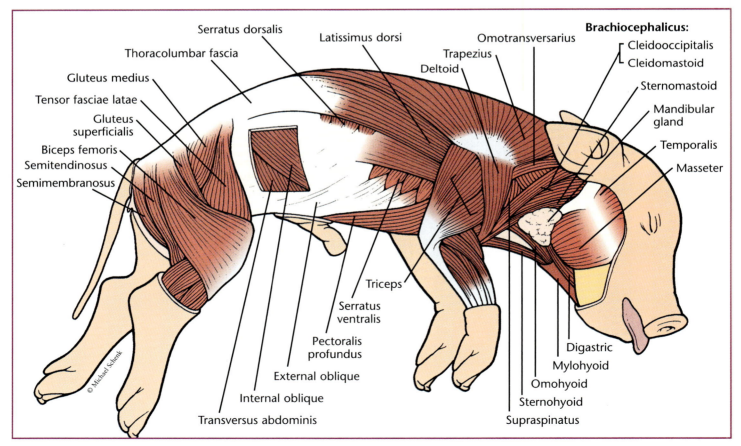

FIGURE 3.5　*Illustration of full lateral view depicting superficial muscles (and deep muscles of the abdominal region).*

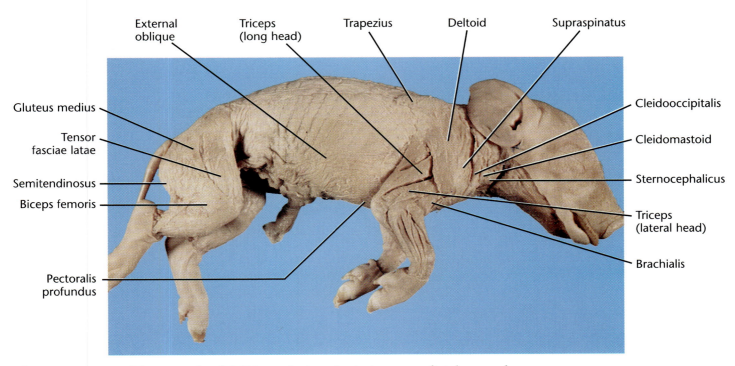

FIGURE 3.6 *Photograph of full lateral view depicting superficial musculature.*

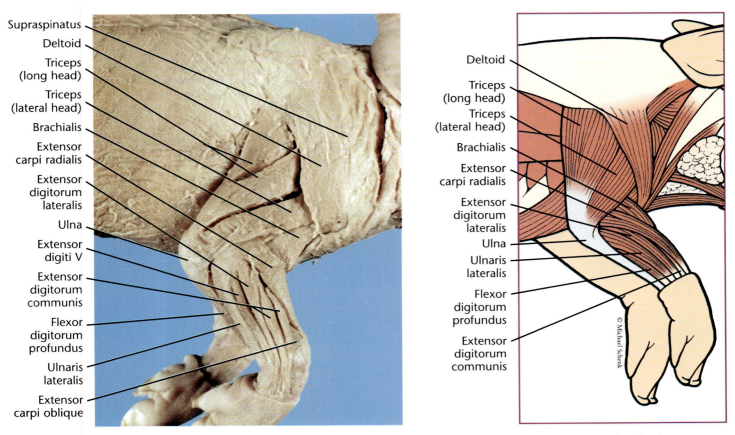

FIGURE 3.7 *Photograph and illustration of lateral view depicting selected muscles of the forelimb.*

scapula. The **deltoid** is a fairly large muscle running from the mid-back (near the **trapezius**), across the front of the shoulder to the junction of the elbow. The **serratus ventralis** is a large, fan-shaped muscle that runs along the ribs. In addition to moving the scapula, this muscle supports a large portion of the weight of the trunk of the body. The **pectoralis superficialis** is located on the ventral side of the chest, cranial to the portion of the pectoralis profundus that crosses the chest. This muscle adducts the forelimb of the pig. The **triceps** (actually a group of several muscles) is located perpendicular to the deltoid, running from the scapula to the back of the elbow. The **brachialis** arises from the humerus and runs underneath the biceps muscle on the forelimb. One of the larger muscles along the lateral side of the pig is the **latissimus dorsi**. This broad muscle attaches to the humerus and aids in retracting the forelimb.

Examine the lateral side of the forelimb on your pig (Figures 3.5–3.7). Notice there is a group of long muscles running from the elbow to the toes. These are the extensor muscles which extend the wrist and digits in the pig. The most medial of these extensor muscles is the **extensor carpi radialis**. The most lateral extensor is the **ulnaris lateralis**. The common extensor, or **extensor digitorum communis** as it is named, lies along the lateral side of the extensor carpi radialis. Lying just to the lateral side of the extensor digitorum communis is the **extensor digitorum lateralis**. Together these four extensors act primarily on the carpus and phalanges to move the foot and forelimb.

TABLE 3.3   *Superficial muscles of the pectoral region and forelimb. Refer to Figures 3.4–3.8.*

Muscle Name	Action
Pectoralis profundus	Adducts forelimb.
Omotransversarius	Assists in advancing the forelimb.
Deltoid	Flexes shoulder; abducts forelimb.
Trapezius	Elevates scapula and draws scapula laterally.
Serratus ventralis	Pulls scapula caudally and downward.
Pectoralis superficialis	Adducts forelimb.
Triceps	Extends forelimb.
Brachialis	Flexes forelimb.
Latissimus dorsi	Flexes shoulder; moves forelimb dorsally and caudally.
Extensor carpi radialis	Extends and flexes the carpus (wrist).
Ulnaris lateralis (extensor)	Extends and flexes the carpus (wrist).
Extensor digitorum communis	Extends the joints of principle digits.
Extensor digitorum lateralis	Extends the fourth digit.
Flexor carpi radialis	Flexes the carpus (wrist) and digits.
Flexor carpi ulnaris	Flexes the carpus (wrist).
Flexor digitorum profundus	Flexes the carpus (wrist) and digits.
Flexor digitorum superficialis	Flexes the proximal and middle joints of the digits.

The flexors of the forelimb and foot are also grouped together but occur on the medial side of the forelimb (Figure 3.8). The **flexor carpi radialis** is the most medial of the flexors and runs directly to the radius (hence its name). The **flexor carpi ulnaris** is the most lateral and extends to the point of the elbow. In between these two muscles lies the **flexor digitorum profundus** (next to the flexor carpi radialis) and the **flexor digitorum superficialis** (next to the flexor carpi ulnaris). These four flexors work antagonistically to the extensors to flex the carpus and phalanges of the pig.

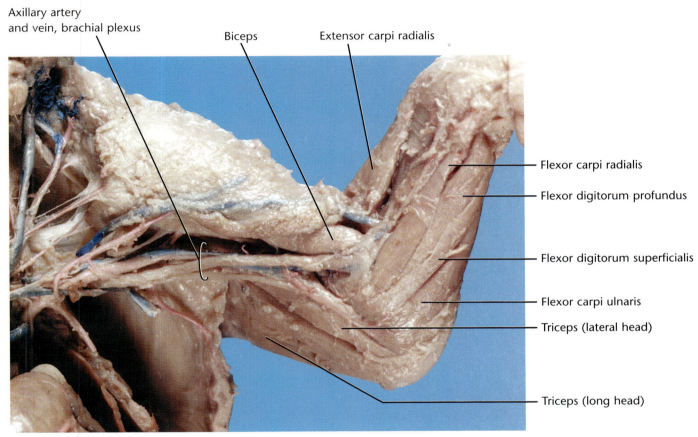

Axillary artery and vein, brachial plexus

Biceps

Extensor carpi radialis

Flexor carpi radialis

Flexor digitorum profundus

Flexor digitorum superficialis

Flexor carpi ulnaris

Triceps (lateral head)

Triceps (long head)

FIGURE 3.8   *Photograph of medial side of forelimb depicting superficial musculature.*

Deep Musculature

> *On the ventral side of the forelimb, cut through the pectoralis superficialis and the pectoralis profundus and reflect them back to expose the underlying deep muscles of the pectoral region and forelimb.*

Underneath the superficial pectoral muscles, locate the **rectus thoracis** and the **rectus abdominis** running along the median plane of the body (Figure 3.3). Lying laterally to the rectus thoracis is the **scalenus**, a muscle that pulls the ribs cranially and aids in respiration. Moving toward the armpit region, locate the **teres major** and the **subscapularis**. These two muscles lie alongside one another and may be difficult to tell apart. The median head of the **triceps** runs along the humerus from the shoulder blade to the point of the elbow. The **coraco-brachialis** runs along this same length, but buries deep inside the forelimb about halfway down the length of the upper forelimb. The **biceps** is the muscle of this region with which we are perhaps most familiar in humans; however, in the pig this muscle is not very prominent. It is located along the caudal portion of the forelimb and serves to flex the forelimb.

TABLE 3.4   *Deep muscles of the pectoral region and forelimb. Refer to Figure 3.3.*

Muscle Name	Action
Rectus thoracis (not depicted)	Enlarges rib cage (assists in respiration).
Rectus abdominis (not depicted)	Contracts the length of the torso.
Scalenus (not depicted)	Pulls ribs cranially (assists in respiration).
Teres major	Flexes shoulder; adducts forelimb.
Subscapularis	Braces shoulder; adducts forelimb.
Triceps	Extends forelimb.
Coracobrachialis	Braces medial aspect of shoulder.
Biceps	Flexes forelimb.

The Abdomen

Position your pig so that the muscles on the ventral side of the abdomen are in view. Use scissors (or a scalpel) to make very shallow incisions through a small portion of the outermost muscle layer of the abdomen. This will allow you to view both the superficial and deep musculature of the trunk region.

The outermost abdominal muscle layer is comprised of the **external obliques** which compress the abdomen and help flex the trunk (Figure 3.5). The muscle fibers of these muscles run diagonally across the abdomen at an oblique angle to the torso, and their name is derived from this arrangement. Underneath this layer you will find the **internal obliques**. Their muscle fibers run at a ninety degree angle to those of the external obliques. The innermost layer of abdominal muscles runs horizontally across the trunk (perpendicular to the long axis of the body) and is comprised of the **transversus abdominis muscles**. Lying dorsal to the external obliques is the **serratus dorsalis**. This muscle helps to raise the rib cage and enlarges the thoracic cavity, aiding in respiration.

TABLE 3.5 *Superficial and deep muscles of the abdominal region. Refer to Figure 3.5.*

Muscle Name	Action
External oblique	Compresses abdomen and flexes trunk.
Internal oblique	Compresses abdomen and flexes trunk.
Transversus abdominis	Compresses abdomen and flexes trunk.
Serratus dorsalis	Raises ribs; enlarges thoracic cavity.

The Pelvic Region and Hindlimb

Superficial Musculature

On the lateral side of the hindlimb there are seven major muscles (Figures 3.5 and 3.9). The most cranial is the **tensor fasciae latae**, which extends the hindlimb. This muscle runs from the crest of the ilium and attaches to the front of the knee. Moving caudally, the next muscle seen is the **rectus femoris**, another muscle which extends the hindlimb. Next, identify the **gluteus medius**, one of the larger muscles in the upper thigh region. In addition to extending the leg, this muscle also abducts the thigh. Moving farther back, locate the **gluteus superficialis**. This small muscle lies alongside the largest muscle of the thigh, the **biceps femoris**. Caudal to the biceps femoris are two more muscles that extend the thigh and hip. They are the **semitendinosus** and the **semimembranosus**. On the distal portion of the hindlimb, locate the two extensor muscles of the hindlimb, the **extensor digitorum longus** and the **extensor digitorum quarti and quinti**. These muscles extend the tarsus joint (ankle) and phalanges of the hindlimb.

On the medial side of the hindlimb, the most cranial thigh muscle is the **sartorius** which adducts the thigh and extends the hindlimb (Figures 3.10 and 3.11). The largest muscle on the medial side of the thigh is the **gracilis**, which also adducts the thigh. Further down on the distal portion of the hindlimb, identify the **tibialis anterior**, the most cranial of the flexor muscles of the hindlimb. Next, identify the **tibialis posterior**, another flexor of the hindlimb located just caudal to the tibia. Behind these two muscles, locate the **flexor digitorum longus** and **flexor hallicus**, the third group of flexor muscles. The **gastrocnemius** and **soleus** comprise the calf muscles in the pig which extend the hindfoot. These are found on the inner portion of the leg, behind the knee joint.

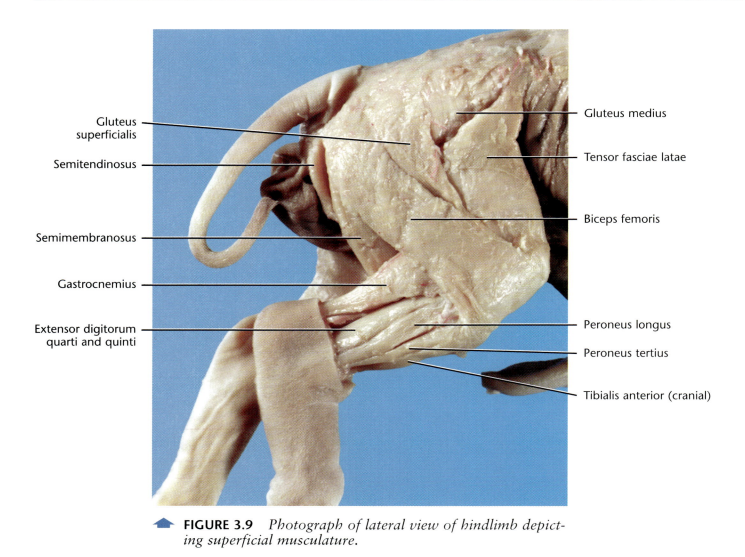

Gluteus superficialis

Semitendinosus

Semimembranosus

Gastrocnemius

Extensor digitorum quarti and quinti

Gluteus medius

Tensor fasciae latae

Biceps femoris

Peroneus longus

Peroneus tertius

Tibialis anterior (cranial)

FIGURE 3.9   *Photograph of lateral view of hindlimb depicting superficial musculature.*

TABLE 3.6   *Superficial muscles of the pelvic region and hindlimb — lateral view. Refer to Figures 3.5, 3.9, 3.11, and 3.12.*

Muscle Name	Action
Tensor fasciae latae	Extends hindlimb.
Rectus femoris	Extends hindlimb.
Gluteus medius	Abducts and extends thigh.
Gluteus superficialis	Abducts thigh.
Biceps femoris	Abducts thigh; flexes hindlimb.
Semitendinosus	Extends thigh and flexes hindlimb.
Semimembranosus	Extends hip and adducts hindlimb.
Extensor digitorum longus	Extends tarsus (ankle) and phalanges.
Extensor digitorum quarti and quinti	Extends hindfoot.

Deep Musculature

Lay your pig on its dorsal side to gain access to the medial side of the thigh region. You may need to tie the hindlimbs "open" with string or pin them down to keep them apart while dissecting the muscles of this region. Cut through the gracilis and reflect it back to expose the underlying deep musculature of the medial side of the hindlimb.

The most cranial of the deep muscles on the medial side of the hindlimb is the **iliacus,** which flexes the hip and rotates the thigh (Figures 3.10 and 3.11). This muscle, along with the **psoas major** (which lies alongside the iliacus), originates from the vertebral column and inserts along the femur near the knee joint. Another deep flexor of the hip is the **pectineus** muscle, which lies next to the sartorius. Moving cranially, the next muscle visible is the **adductor** which, as its name implies, adducts the thigh. The last deep muscle of the thigh is the **vastus medialis** which originates from the medial aspect of the femur and inserts upon the juncture of the tibia (near the patella). From this position the vastus medialis is capable of extending the hindlimb.

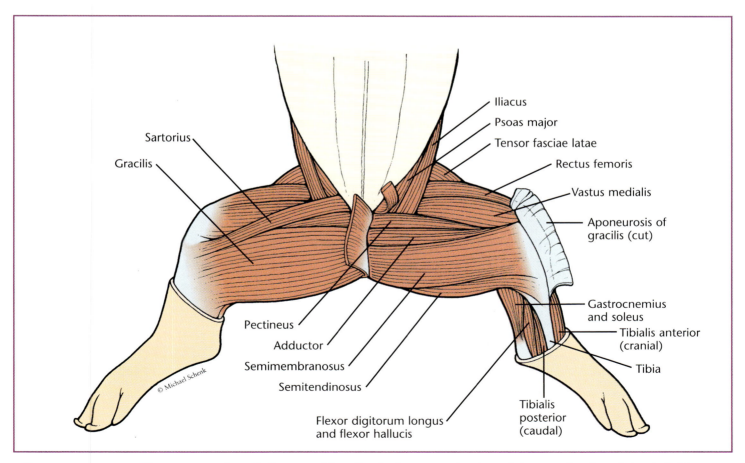

FIGURE 3.10   *Illustration of medial view of hindlimb depicting (pig's right) superficial and (pig's left) selected deep muscles.*

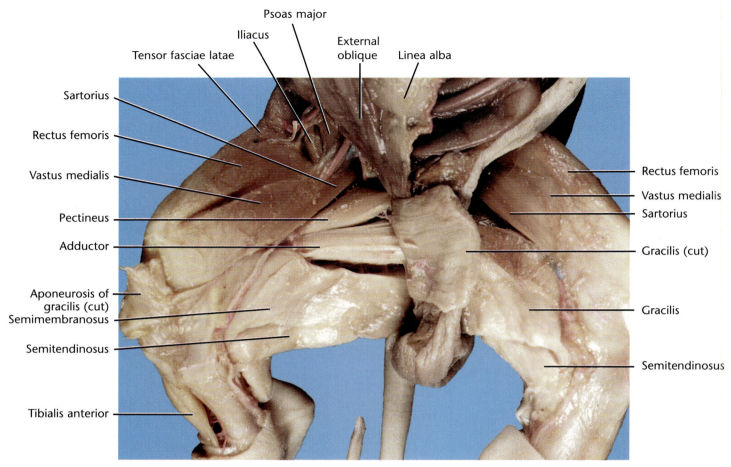

FIGURE 3.11   *Photograph of medial view of hindlimb depicting selected (pig's right) superficial and (pig's left) deep muscles.*

TABLE 3.7   *Superficial muscles of the pelvic region and hindlimb — medial view. Refer to Figures 3.10 and 3.11.*

Muscle Name	Action
Sartorius	Extends hindlimb; adducts thigh.
Gracilis	Adducts thigh.
Tibialis anterior	Flexes hindfoot.
Tibialis posterior	Flexes and inverts hindfoot.
Flexor digitorum longus and flexor hallicus	Flex hindfoot.
Gastrocnemius and soleus	Extend hindfoot.

TABLE 3.8   *Deep muscles of the pelvic region and hindlimb — medial surface. Refer to Figures 3.10 and 3.11.*

Muscle Name	Action
Iliacus	Flexes hip and rotates thigh.
Psoas major	Flexes hip and rotates thigh.
Pectineus	Flexes hip and adducts thigh.
Adductor	Adducts thigh.
Vastus medialis	Extends hindlimb.

On the lateral side of the hindlimb, cut through the gluteus superficialis, biceps femoris and tensor fasciae latae.

Once the biceps femoris is cut, you will be able to see the large **vastus lateralis** which extends the hindlimb (Figures 3.12 and 3.13). This prominent muscle (like its medial counterpart) extends from the femur to the tibia near the knee. Caudal to this muscle is the **quadratus femoris**, one of the quadricep muscles of the thigh which extend the hip and hindlimb. Cranial to the vastus lateralis is the small **gluteus profundus**, almost completely covered by the gluteus medius. This muscle abducts the hindlimb. Further down on the distal portion of the hindlimb, you should locate two muscles that were exposed when the biceps femoris was cut and reflected back. These are the **peroneus longus** and smaller **peroneus tertius,** flexors of the tarsus located along the cranial aspect of the tibia.

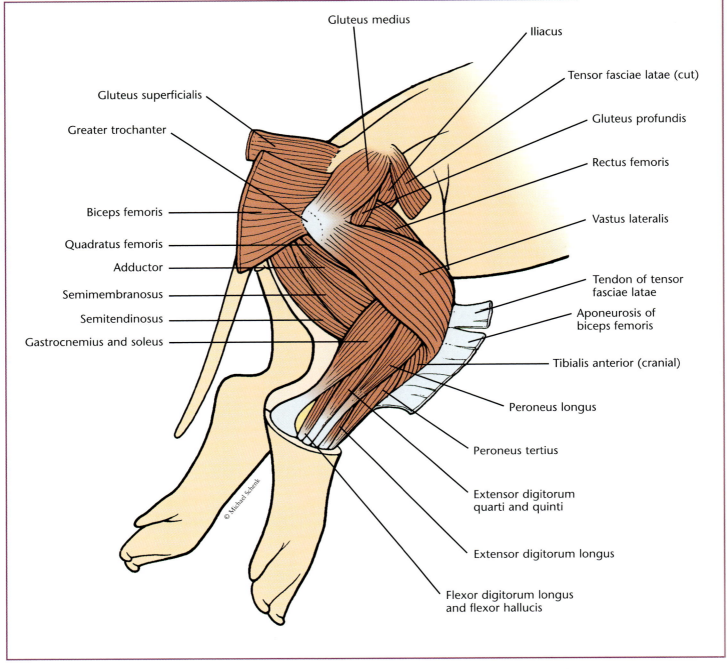

FIGURE 3.12   *Illustration of lateral view of hindlimb depicting selected deep muscles.*

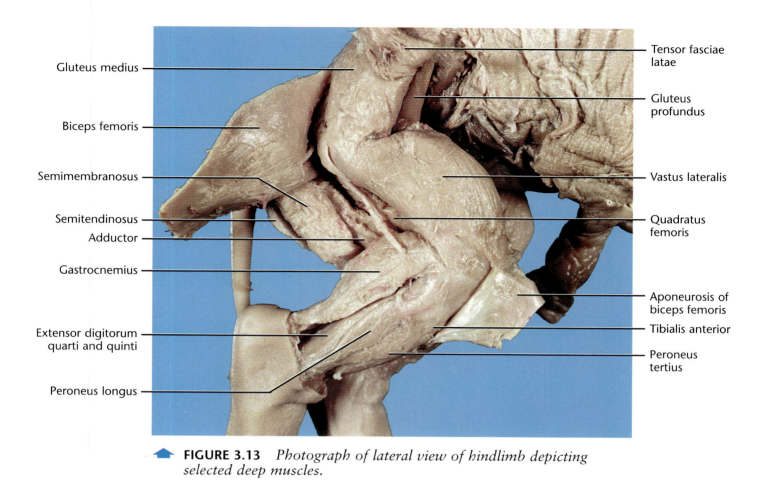

Gluteus medius

Biceps femoris

Semimembranosus

Semitendinosus

Adductor

Gastrocnemius

Extensor digitorum
quarti and quinti

Peroneus longus

Tensor fasciae
latae

Gluteus
profundus

Vastus lateralis

Quadratus
femoris

Aponeurosis of
biceps femoris

Tibialis anterior

Peroneus
tertius

FIGURE 3.13 *Photograph of lateral view of hindlimb depicting selected deep muscles.*

TABLE 3.9 *Deep muscles of the pelvic region and hindlimb — lateral surface. Refer to Figures 3.12 and 3.13.*

Muscle Name	Action
Vastus lateralis	Extends hindlimb.
Quadratus femoris	Extends hip.
Gluteus profundus	Abducts thigh and rotates it medially.
Peroneus longus	Flexes hindfoot; abducts and everts.
Peroneus tertius	Flexes and everts hindfoot.

CHAPTER 4

DIGESTIVE SYSTEM

Laboratory Objectives

After completing this chapter, you should be able to:

1. Identify the major digestive organs of the fetal pig.
2. Describe the functions of all indicated structures.
3. Identify the digestive enzymes produced by the digestive glands and describe their functions.
4. Recognize the microanatomy of the digestive organs.

The digestive system is responsible for mechanically and chemically breaking down food into smaller, usable compounds and then transporting those nutrients into the bloodstream through which they are transported to the individual cells of the body. This process provides the crucial raw materials and energy for all metabolic processes carried out by the organism.

Head, Neck and Oral Cavity

Lay your pig on its side and observe the salivary glands in the neck region that were exposed when you removed the skin around the neck to view the musculature earlier. These structures lay just below the skin and may have been destroyed when dissecting the muscles. If so, use the other side of the neck and carefully remove the skin from around this region to expose these glands (Figures 4.1 and 4.2).

Locate the **parotid gland,** the largest of the salivary glands in the pig (Figure 4.2). Emanating from the rostral (toward the nose) end of this gland, you should be able to uncover the **parotid duct** which carries the digestive enzymes from the parotid gland into the oral cavity where they mix with food. Underneath the parotid gland there is a small, oval-shaped gland known as the **submaxillary gland.** The third salivary gland to identify is the **sublingual gland** which is flat and narrow and lies underneath the skin alongside the tongue. Saliva plays an important role in the digestive process of mammals by lubricating the food and starting the digestive reactions. In humans and a few other

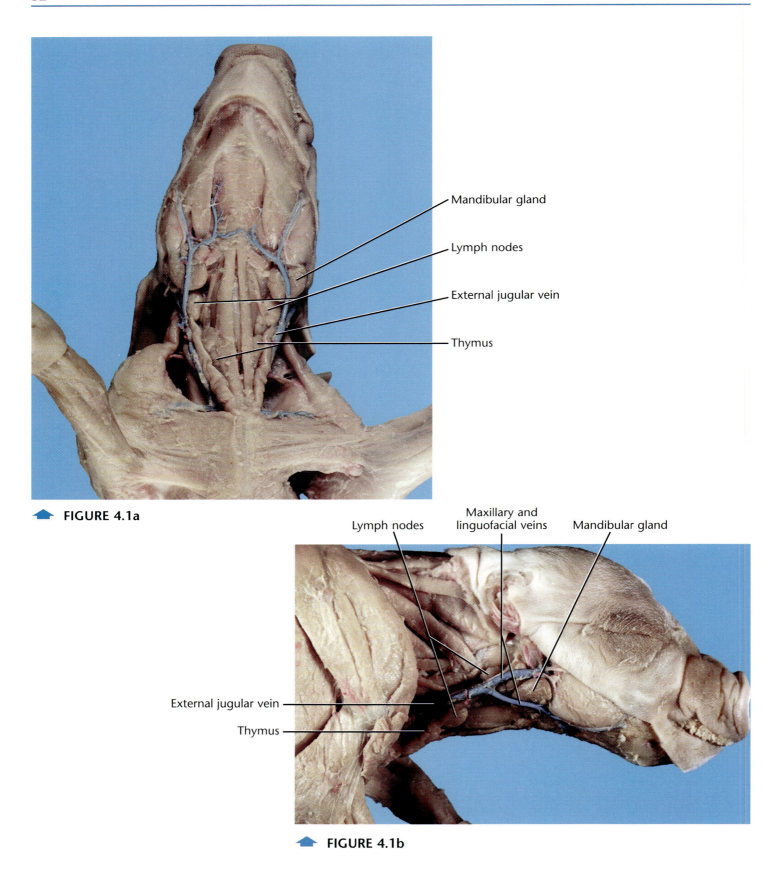

FIGURE 4.1a

FIGURE 4.1b

FIGURE 4.1   *Photographs of head/neck region showing salivary glands, (a) ventral view and (b) lateral view.*

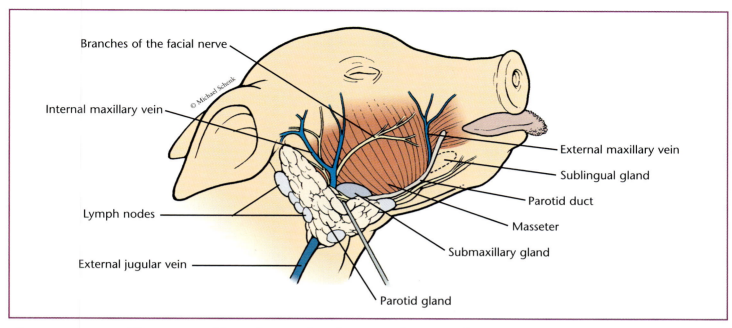

© Michael Schenk

Branches of the facial nerve

Internal maxillary vein

Lymph nodes

External jugular vein

Parotid gland

External maxillary vein

Sublingual gland

Parotid duct

Masseter

Submaxillary gland

FIGURE 4.2 *Illustration of lateral view of neck region showing salivary glands.*

mammals, **amylase** is released by these glands and is responsible for the break-down of starches and glycogen.

> *Using your scalpel (or scissors) make a cut from the corner of the mouth toward the ear on each side of the pig's head. This will extend the opening of the mouth and allow you to view the structures associated with the oral cavity. Don't be afraid to cut too far; usually if you cannot see the structures indicated in the diagram (Figure 4.3), you have not cut far enough.*

If your pig is sufficiently mature, there will be **teeth** protruding from the roof of the mouth (Figure 4.3). In this region of the mouth, the roof is comprised of a bony **hard palate** separating the oral cavity from the nasal cavity above. The **soft palate** is a continuation caudally from the hard palate. This structure is more fleshy in its consistency. The advent of the complete secondary palate allowed mammals to eat and breath simultaneously — one characteristic that allows mammals to have such a high metabolic rate, making endothermy possible. Just caudal to the soft palate is the opening to the **nasopharynx**. This chamber leads rostrally to the external nares. The opening to the **esophagus** should be visible. Next, locate the **glottis**, the opening into the larynx. When the pig swallows, this opening is protected by a thin flap of cartilage called the **epiglottis**. Slowly close the oral cavity and notice how the glottis and epiglottis meet up with the opening to the nasopharynx. On the lower jaw, locate the **tongue**. Notice that there are small bumps near the tip and base of the tongue. These are called **papillae**, and they help the pig manipulate food in its mouth.

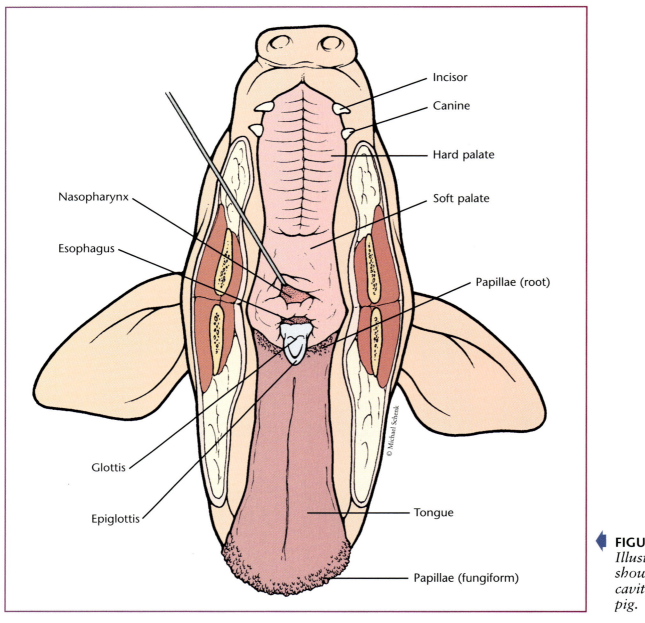

Incisor

Canine

Hard palate

Soft palate

Nasopharynx

Esophagus

Papillae (root)

© Michael Schenk

Glottis

Epiglottis

Tongue

Papillae (fungiform)

FIGURE 4.3
Illustration showing oral cavity of fetal pig.

Abdominal Cavity

Use Figure 3.2 as a guide to make the necessary cuts through the muscle layers to expose the digestive organs in the abdominal cavity. The muscle layers are quite thin and you should take care not to cut too deeply through them. Many preserved specimens contain large amounts of liquid preservatives in the body cavities. You may wish to drain these out of your pig, or use a paper towel or sponge to remove them, before proceeding with the identification of the digestive organs.

A thin muscular layer (the diaphragm) separates the upper thoracic cavity from the lower abdominal cavity (Figures 4.4 and 4.5). The food that is swallowed passes down the **esophagus** and into the **stomach.** The stomach lies on the left side of the pig underneath the large, dark liver. It is a J-shaped sac that is responsible for storing large quantities of food. This prevents mammals from having to eat constantly. The stomach releases several chemical compounds that assist the digestive process. **Hydrochloric acid** and **pepsinogen** are two digestive compounds released by the stomach.

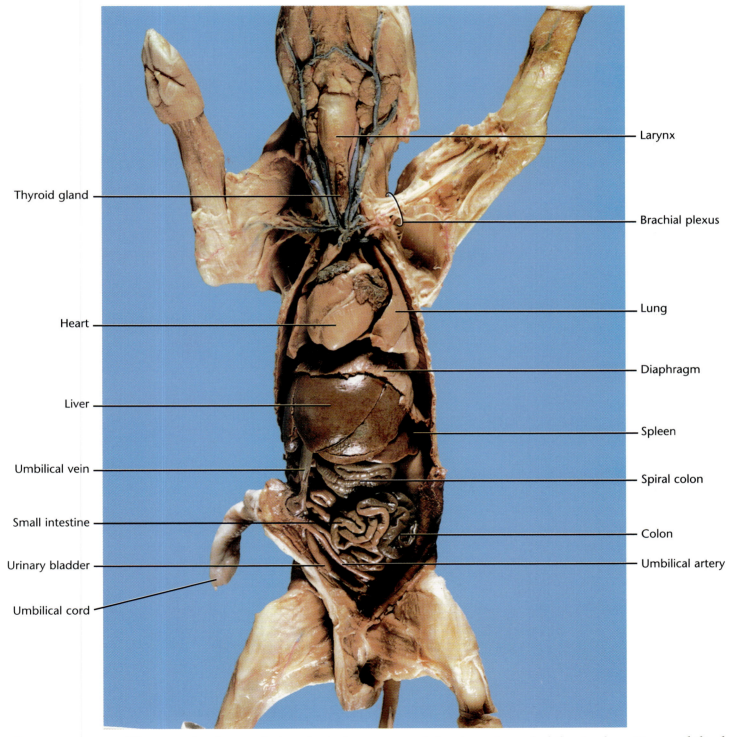

FIGURE 4.4   *Photograph of ventral view depicting organs of the thoracic and abdominal cavities; umbilical vein still intact.*

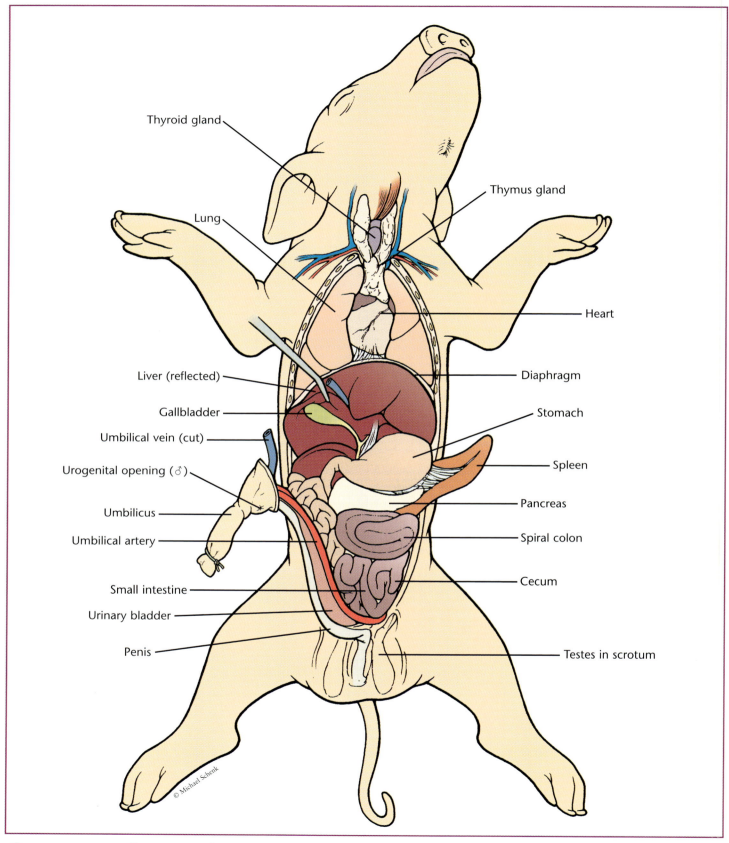

FIGURE 4.5   *Illustration of ventral view depicting organs of the thoracic and abdominal cavities; umbilical vein cut.*

Make an incision along the lateral margin of the stomach to expose its interior.

Notice that there are small folds inside the stomach. These are called **rugae**, and they help churn the food and mix it with chemical secretions. The stomach empties its contents into the **duodenum** — the first portion of the small intestine. At this point, several accessory glands empty digestive fluids into the duodenum. Locate the **liver**, the largest organ in the abdominal region. The liver produces bile which is stored in the **gallbladder**. The gallbladder is located on the underside of the right lobe of the liver (Figure 4.6). **Bile** contains no digestive enzymes, but it does contain bile salts which assist in the breakdown of

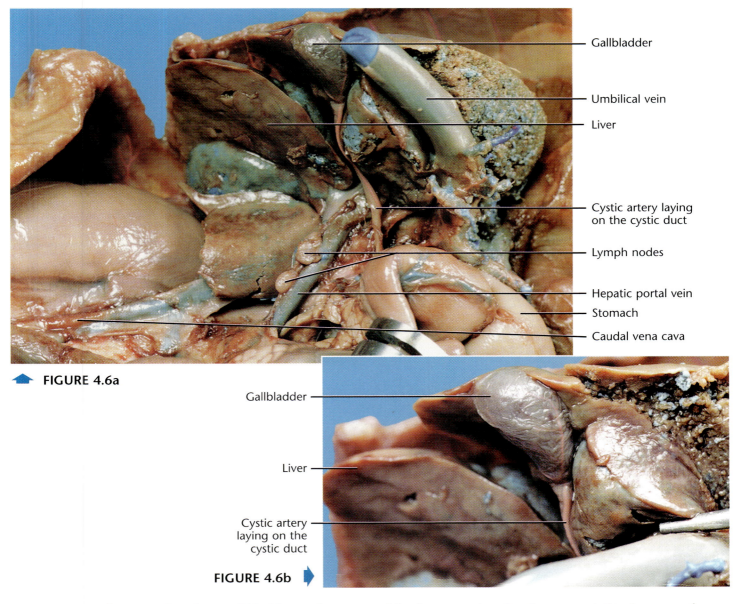

Gallbladder

Umbilical vein

Liver

Cystic artery laying on the cystic duct

Lymph nodes

Hepatic portal vein

Stomach

Caudal vena cava

◆ **FIGURE 4.6a**

Gallbladder

Liver

Cystic artery laying on the cystic duct

FIGURE 4.6b ▶

FIGURE 4.6   *Photographs of (a) gallbladder and common bile duct entering duodenum and (b) close-up of gallbladder showing arterial supply (with liver partially removed in both photos).*

fats. The bile is released directly into the **cystic duct** which carries the bile into the **common bile duct** and then into the duodenum. Find the **pancreas**, a whitish-yellow, elongated, granular organ which is imbedded in the mesenteries that support the stomach (Figure 4.7). The pancreas is actually composed of two lobes, a left lobe which runs transversely across the body and a smaller, right lobe which runs longitudinally along the length of the duodenum (Figure 4.8). The pancreas produces several kinds of digestive enzymes and hormones. The digestive enzymes empty into the duodenum via the small **pancreatic duct** and **accessory duct** (Figure 4.9). The duodenum receives the partially digested foodstuffs and enzyme mix, known as **chyme**, from the stomach and is primarily responsible for the final stages of enzymatic digestion. Food passes next into the **jejunum** (Figure 4.10). This region of the small intestine is highly convoluted and tightly bound together by **mesentery**. Absorption of nutrients and water occurs along the length of the jejunum. Chyme then passes into the distal

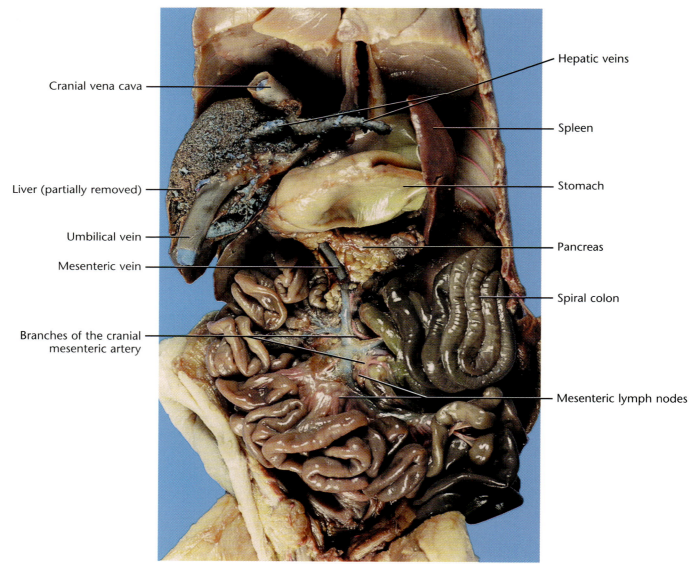

FIGURE 4.7 *Photograph of ventral view of abdominal cavity with organs displaced to expose underlying structures (liver partially removed).*

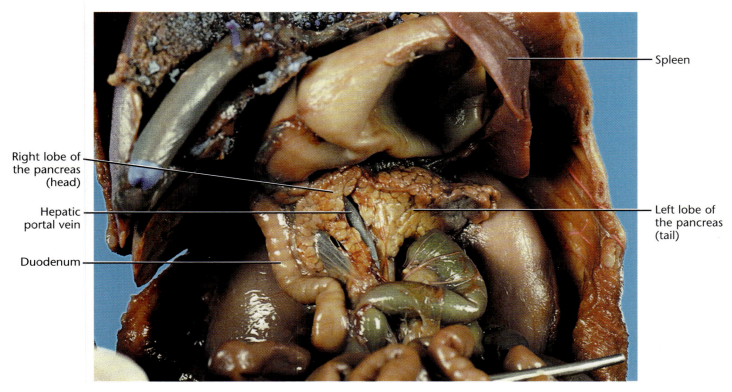

Spleen

Right lobe of
the pancreas
(head)

Hepatic
portal vein

Duodenum

Left lobe of
the pancreas
(tail)

◀ **FIGURE 4.8** *Close-up of pancreas depicting left and right lobes on opposite sides of the hepatic portal vein.*

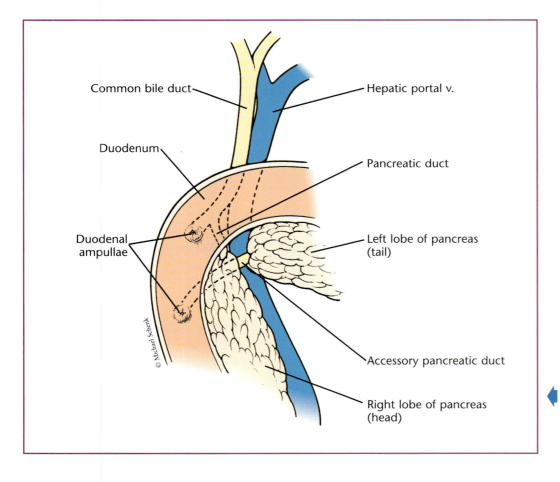

Common bile duct

Hepatic portal v.

Duodenum

Pancreatic duct

Duodenal
ampullae

Left lobe of pancreas
(tail)

© Michael Schenk

Accessory pancreatic duct

Right lobe of pancreas
(head)

◀ **FIGURE 4.9** *Illustration of pancreatic ducts and their connections to the duodenum.*

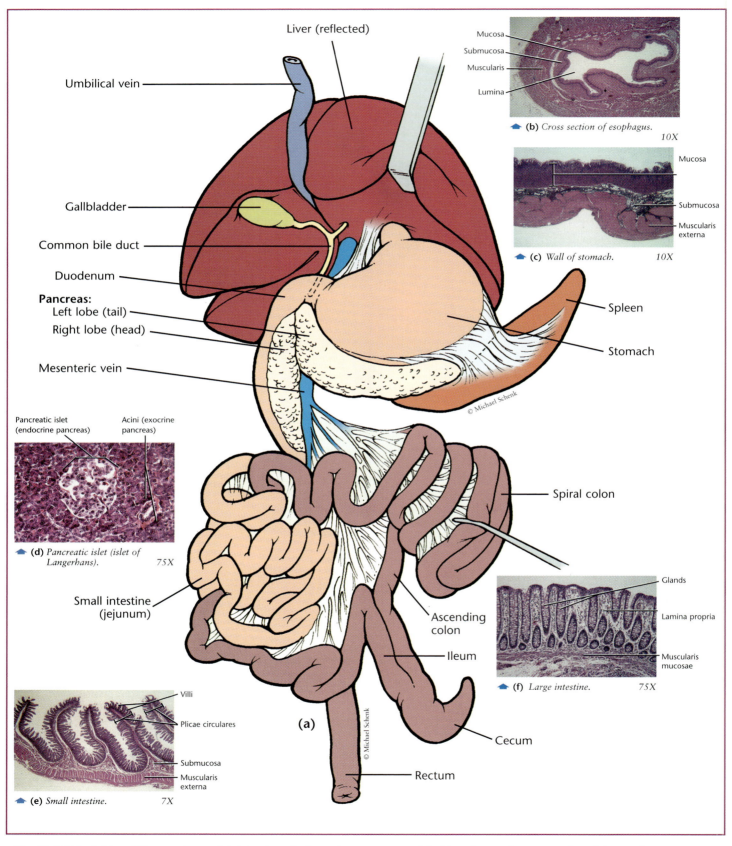

Liver (reflected)

Umbilical vein

Mucosa
Submucosa
Muscularis
Lumina

▲ **(b)** *Cross section of esophagus.*   *10X*

Gallbladder

Common bile duct

Duodenum

Mucosa

Submucosa

Muscularis externa

▲ **(c)** *Wall of stomach.*   *10X*

Pancreas:
Left lobe (tail)
Right lobe (head)

Spleen

Mesenteric vein

Stomach

© Michael Schenk

Pancreatic islet
(endocrine pancreas)

Acini (exocrine pancreas)

Spiral colon

▲ **(d)** *Pancreatic islet (islet of Langerhans).*   *75X*

Small intestine
(jejunum)

Ascending colon

Glands

Lamina propria

Muscularis mucosae

Ileum

▲ **(f)** *Large intestine.*   *75X*

(a)

© Michael Schenk

Villi

Plicae circulares

Cecum

Submucosa
Muscularis externa

Rectum

▲ **(e)** *Small intestine.*   *7X*

▲ **FIGURE 4.10**   *Illustration of (a) isolated digestive system with histology photographs, (b) esophagus, (c) wall of stomach, (d) pancreas, (e) small intestine, (f) large intestine.*

portion of the small intestine known as the **ileum** where further nutrient absorption and water reabsorption occur. At the juncture of the small intestine and the tightly-coiled colon there is a short, blind-ended pocket of the intestine known as the **cecum**. In carnivores and omnivores, the cecum is very small and does not play a large role in digestion. However, in herbivores, the cecum is typically quite large and serves as a fermentation chamber where symbiotic bacteria break down cellulose and other plant matter. The mixture passes next through the **ascending colon** into the tightly-coiled **spiral colon** which is primarily responsible for reabsorption of water. Locate the descending portion of the colon that runs along the dorsal aspect of the abdominal cavity. This portion is called the **rectum**. The colon and rectum permit the pig to conserve valuable water and produce a dry feces. From the beginning of the digestive process, fluid-based chemicals have been mixed in with the food. At this point, most usable nutrients have been dissolved and absorbed by the duodenum, jejunum and ileum, and the water that was previously added is now reabsorbed. The undigested food particles (feces) are finally egested from the pig through the **anus**.

TABLE 4.1 *Digestive organs in the fetal pig and their functions. Structures denoted with an asterisk (*) are accessory digestive organs. Food does not pass directly into these accessory organs; however, they do play a major role in the digestive process.*

Organ/Structure	Function
Teeth	Mechanically breakdown food.
Salivary glands	Secrete digestive enzymes (e.g. amylase) to begin chemical breakdown of foods and lubricate food for swallowing.
Esophagus	Transports food to stomach.
Stomach	Produces hydrochloric acid and pepsinogen that aid in the chemical breakdown of food.
Duodenum	Receives chyme from the stomach along with secretory enzymes from the gallbladder and pancreas.
Liver*	Produces bile, converts glucose to glycogen for storage, detoxifies many constituents of the absorbed digested compounds.
Gallbladder*	Stores bile for breakdown of fats.
Pancreas*	Produces digestive enzymes and delivers them through pancreatic duct to duodenum.
Jejunum	Responsible for majority of nutrient absorption and reabsorption of water.
Ileum	Continues process of nutrient absorption and reabsorption of water.
Cecum	Has a reduced appearance and function in carnivores and omnivores. (In herbivores, this structure is quite large and contains anaerobic bacteria responsible for fermentation of cellulose and other plant materials.)
Colon	Responsible for reabsorption of water and electrolytes; produces feces.
Rectum	Final site of water reabsorption and feces production.
Anus	Regulates egestion of undigested food (feces) from the body.

CARDIOVASCULAR SYSTEM

Laboratory Objectives

After completing this chapter, you should be able to:

1. Identify the major arteries and veins of the fetal pig.
2. Identify the chambers of the fetal heart.
3. Discuss the function of all indicated structures.
4. Discuss the circulatory pathway of blood in the fetal pig and contrast it to the pathway of blood after birth.

The cardiovascular (or circulatory) system is responsible for transporting nutrients, gases, hormones and metabolic wastes to and from the individual cells of an organism. Organisms like mammals are far too large for all of their individual cells to exchange nutrients, wastes and gases with the external world by simple diffusion. Most cells are buried too deep inside the body to effectively accomplish this task. Thus, some system must be in place to efficiently exchange these products between the outside world and every cell in the organism's body. For this reason, the cardiovascular system is a highly-branched network of vessels that spreads throughout the entire organism.

Thoracic Cavity and Neck Region

Using scissors, extend the incision made earlier through the thoracic region by cutting through the midline of the rib cage to expose the heart and lungs. The heart will be visible in the center of the thoracic cavity, encased in the pericardial membrane.

The Heart

Notice the thin **pericardial membrane** surrounding the heart (Figure 5.1A). This protective sac contains a small amount of lubricating fluid to protect the heart and cushion its movements. On the cranial surface of the pericardial membrane, a portion of the thymus gland will be evident. Do not confuse this with either of the atria of the heart which lie inside the pericardial membrane. The

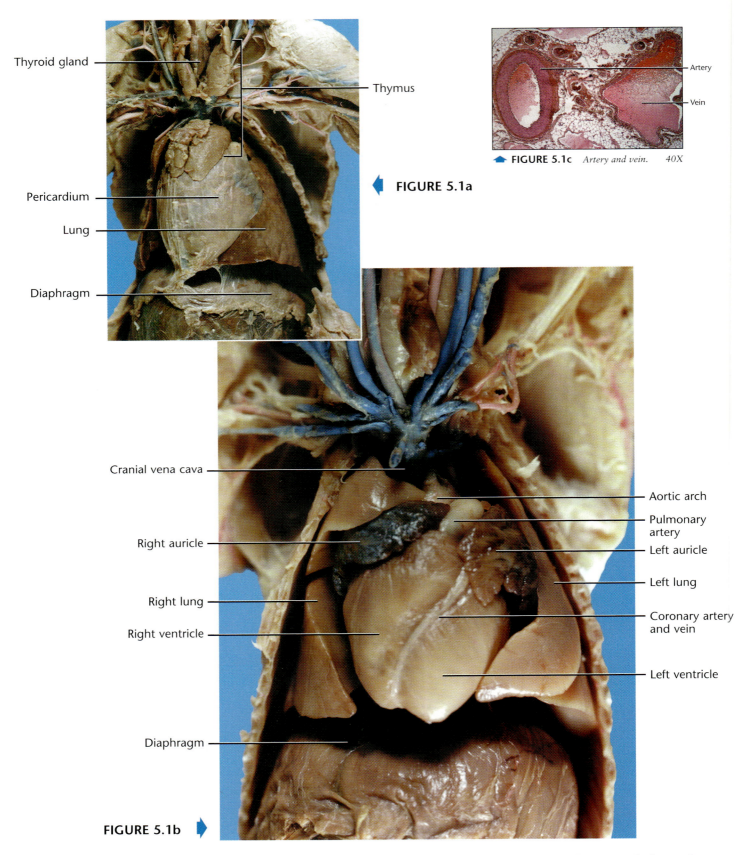

Thyroid gland

Thymus

Artery

Vein

FIGURE 5.1c *Artery and vein.*   40X

FIGURE 5.1a

Pericardium

Lung

Diaphragm

Cranial vena cava

Aortic arch

Pulmonary artery

Right auricle

Left auricle

Right lung

Left lung

Right ventricle

Coronary artery and vein

Left ventricle

Diaphragm

FIGURE 5.1b

FIGURE 5.1 *Photographs of (a) heart surrounded by pericardial membrane, (b) heart with pericardial membrane removed, and (c) histology photograph of cross-section through an artery and vein.*

thymus gland will be discussed in detail in Chapter 9. You should see several arteries and veins emerging from the pericardial membrane. This membrane fits snugly against these arteries and veins and must be carefully removed to fully view these vessels and the other regions of the heart itself.

Carefully remove the pericardial membrane from the heart. Use a teasing needle and forceps to carefully dissect the muscle tissue and fatty tissue away from the major arteries and veins in the neck region. This is a tedious process and will take some time. Use Figure 5.1B as a guide. If your pig has been double-injected with latex, the arteries will appear red and the veins will appear blue. If your pig has not been injected with latex, the arteries will appear whiter and stiffer than the thin, collapsed veins. Remember that arteries are more heavily walled than veins (to accommodate higher blood pressures) and may appear thicker.

Identify the four chambers of the pig heart (Figure 5.5). Caudally there are two large, thick-walled ventricles, the **right ventricle** and the **left ventricle**. These chambers pump blood out of the heart to the lungs and to the rest of the body, respectively. Cranial to the ventricles and somewhat different in color and texture are the **right atrium** and **left atrium**. These chambers receive the blood from the body and the lungs, respectively, and pass it on to the ventricles. Running along the surface of the heart itself, the small **coronary arteries** (Figure 5.2a) should be evident. These vessels supply blood to the heart muscle, insuring that it too receives nutrients and oxygen.

Notice the large veins entering the heart on the right side. These are the **cranial** and **caudal vena cavae** (Figure 5.2a). They bring deoxygenated blood to the right atrium (Figure 5.2b) from the upper and lower portions of the body. On the dorsal surface of the heart, adjacent to the juncture of the vena cavae and the right atrium, there is a small sac-like region of the heart known as the **coronary sinus** (Figure 5.2e). This sinus is responsible for returning deoxygenated blood from the wall of the heart to the right atrium. The most visible artery from the ventral surface leaving the heart is the large **pulmonary artery** (Figure 5.2d) emanating from the right ventricle. In the adult this artery would channel blood from the right ventricle through the right and left pulmonary arteries to the lungs. Follow the pulmonary artery behind the heart and see where it branches into the **right and left pulmonary artery**. Notice that at the base of the pulmonary artery there is a connection between it and the aorta (the large artery leaving the left ventricle). This connection is called the ductus arteriosus (Figure 5.2c), a short linkage found only in the fetus. Lying adjacent to the pulmonary arteries are the **pulmonary veins**, the vessels that, in the adult, return oxygenated blood to the left atrium of the heart.

There are three major differences in the circulatory system of a fetal and adult pig (Figure 5.3). The most obvious is the connection between the fetus and the mother through the **umbilical cord** (Figure 5.3c). This collection of tubes allows oxygen and nutrients to pass from the mother to the fetus while transporting carbon dioxide and metabolic wastes away from the fetus. The other two major differences lie in the structure of the heart. Since the fetus is not breathing with its lungs, the lungs do not oxygenate blood that passes through them. In fact, the most oxygen-rich blood in the fetal circulatory system is that returning to the fetus through the umbilical vein. As a result, only a small fraction of the blood leaving the right ventricle travels through the

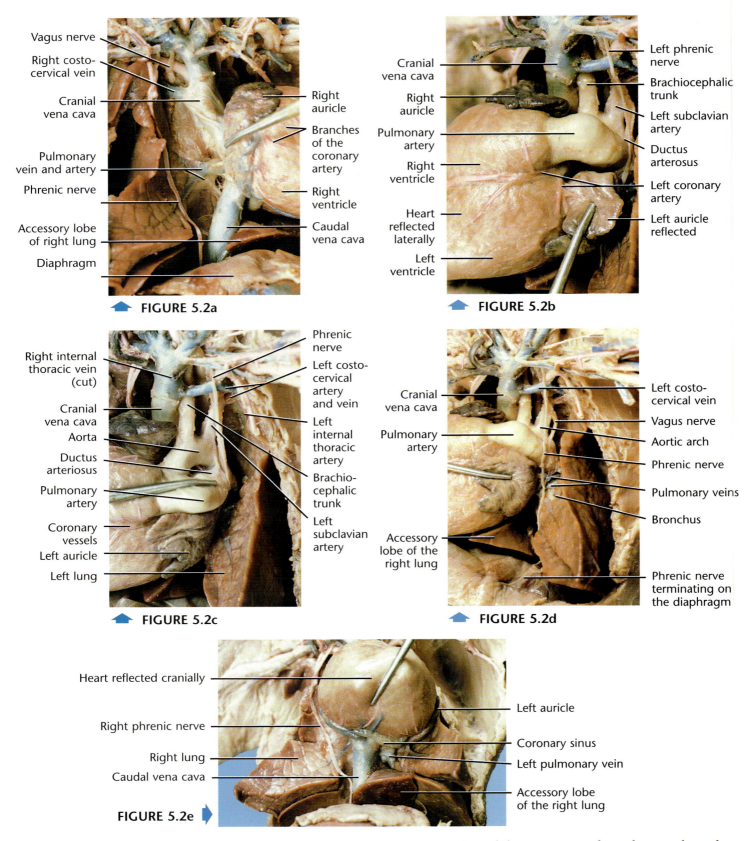

FIGURE 5.2a

Vagus nerve
Right costo-cervical vein
Cranial vena cava
Pulmonary vein and artery
Phrenic nerve
Accessory lobe of right lung
Diaphragm
Right auricle
Branches of the coronary artery
Right ventricle
Caudal vena cava

FIGURE 5.2b

Cranial vena cava
Right auricle
Pulmonary artery
Right ventricle
Heart reflected laterally
Left ventricle
Left phrenic nerve
Brachiocephalic trunk
Left subclavian artery
Ductus arterosus
Left coronary artery
Left auricle reflected

FIGURE 5.2c

Right internal thoracic vein (cut)
Cranial vena cava
Aorta
Ductus arteriosus
Pulmonary artery
Coronary vessels
Left auricle
Left lung
Phrenic nerve
Left costo-cervical artery and vein
Left internal thoracic artery
Brachio-cephalic trunk
Left subclavian artery

FIGURE 5.2d

Cranial vena cava
Pulmonary artery
Accessory lobe of the right lung
Left costo-cervical vein
Vagus nerve
Aortic arch
Phrenic nerve
Pulmonary veins
Bronchus
Phrenic nerve terminating on the diaphragm

FIGURE 5.2e

Heart reflected cranially
Right phrenic nerve
Right lung
Caudal vena cava
Left auricle
Coronary sinus
Left pulmonary vein
Accessory lobe of the right lung

FIGURE 5.2   *Composite photograph of heart anatomy, (a) cranial and caudal vena cavae, (b) right auricle and coronary arteries, (c) aorta, pulmonary trunk, and ductus arteriosus, (d) pulmonary arteries and veins and (e) coronary sinus and caudal vena cava.*

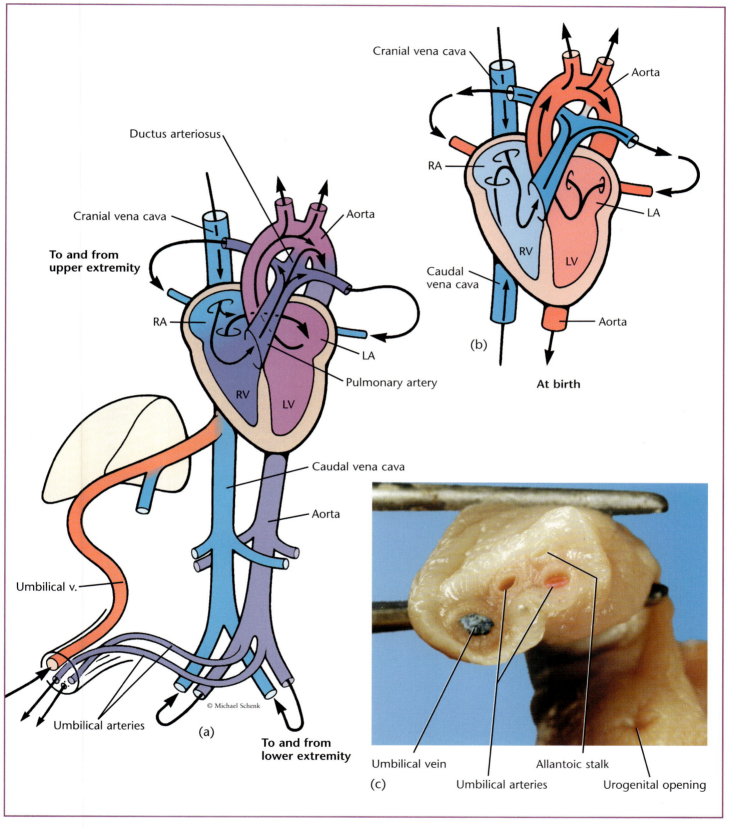

Ductus arteriosus

Cranial vena cava

**To and from
upper extremity**

RA

Aorta

LA

Pulmonary artery

RV

LV

Caudal vena cava

Aorta

Umbilical v.

© Michael Schenk

Umbilical arteries

(a)

**To and from
lower extremity**

Cranial vena cava

Aorta

RA

LA

RV

LV

Caudal
vena cava

Aorta

(b)

At birth

Umbilical vein

Umbilical arteries

Allantoic stalk

Urogenital opening

(c)

FIGURE 5.3   *Schematic illustrations of (a) the fetal circulatory pathway and (b) the comparative pathway after birth, with (c) inset photograph of umbilical cord.*

pulmonary arteries to the lungs. The majority is redirected through the **ductus arteriosus,** a connection between the pulmonary artery and the aorta that channels blood into the aorta (Figure 5.4). Another structure inside the heart of the fetus, called the **foramen ovale,** also aids in rerouting blood to bypass the lungs. This opening in the septum between the right and left atria allows blood passing into the right atrium to be channeled into the left atrium and away from the lungs. Both of these adaptations ensure that the majority of oxygenated blood arriving via the umbilical vein is passed through the fetal circulatory system via the aorta, while permitting enough blood to reach the lungs to allow the lung tissue to develop.

Veins of the Thoracic Region

The largest veins in the fetal pig are the **cranial vena cava** and **caudal vena cava,** which converge at the entrance to the right atrium (Figures 5.5 and 5.6). In the

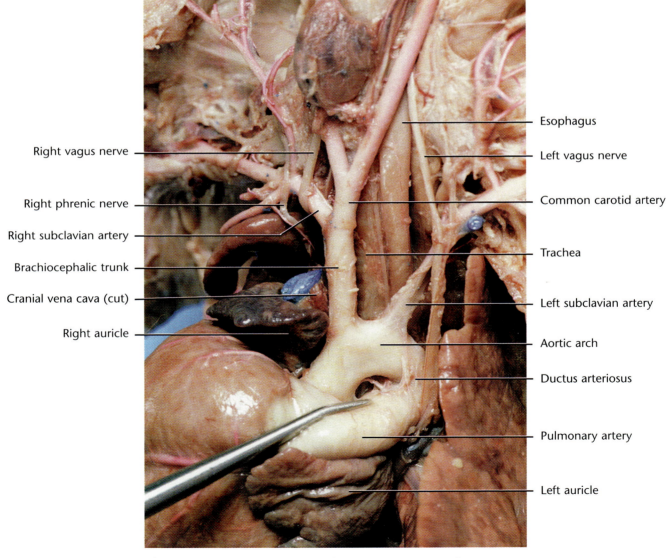

Right vagus nerve
Right phrenic nerve
Right subclavian artery
Brachiocephalic trunk
Cranial vena cava (cut)
Right auricle

Esophagus
Left vagus nerve
Common carotid artery
Trachea
Left subclavian artery
Aortic arch
Ductus arteriosus
Pulmonary artery
Left auricle

FIGURE 5.4   *Close-up of ductus arteriosus depicting connection between pulmonary trunk and the aorta.*

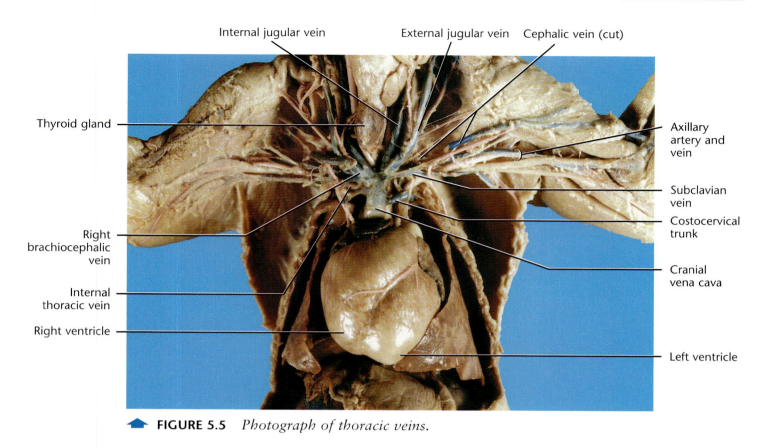

Internal jugular vein   External jugular vein   Cephalic vein (cut)

Thyroid gland

Axillary
artery and
vein

Subclavian
vein

Costocervical
trunk

Right
brachiocephalic
vein

Cranial
vena cava

Internal
thoracic vein

Right ventricle

Left ventricle

FIGURE 5.5   *Photograph of thoracic veins.*

adult, these two thin-walled veins bring deoxygenated blood back to the heart from all parts of the body. Follow the cranial vena cava cranially to its first major branch. Here several smaller veins come together to form the cranial vena cava. Identify the **internal thoracic vein** leading from the arm pit toward the vena cava and heart at a ninety degree angle. The **left and right axillary veins** also lead toward the vena cava at ninety degree angles and bring blood from the forelimbs of the pig. The **subscapular vein** leading from the arm pit and the axillary vein come together to form the **subclavian vein** which dumps blood directly into the cranial vena cava. The third vein returning blood from each forelimb is the **cephalic vein**, the most cranial of the three. The **external jugular veins** lead down into the vena cava from the neck region, along with the **internal jugular veins** running medially alongside the trachea from the head toward the heart. For a short distance the axillary, external jugular, internal jugular and cephalic veins all join together to form the **brachiocephalic vein** which then leads directly to the vena cava. Follow the external jugular vein cranially to the point where it bifurcates into the **linguofacial vein** and the **maxillary vein**. These veins return blood from the lower jaw and the temporal portion of the face, respectively.

Arteries of the Thoracic Region

The first major branch off the aorta is the **brachiocephalic trunk**, which immediately splits into the **right subclavian artery** (which carries blood to the right

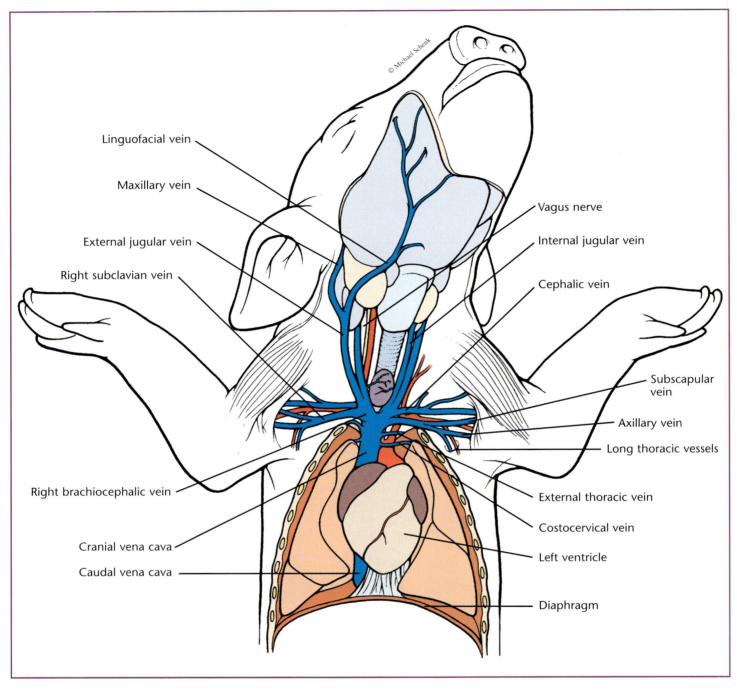

FIGURE 5.6 *Illustration of thoracic veins.*

forelimb and upper portion of the body) and the carotid trunk, off which the **common carotid arteries** branch (which carry blood to the head and brain) (Figures 5.7 and 5.8). Follow the common carotid cranially to the point where it bifurcates into the **external carotid artery** (which runs along the ventral side of the masseter) and the **internal carotid artery** (which embeds underneath the masseter muscle). The **axillary artery** is a continuation of the right subclavian artery which carries blood into the armpit and shoulder region. A small artery may be seen branching off the axillary artery and leading caudally toward the ribs. This is the **internal thoracic artery**. The second major branch off the aorta is the **left subclavian artery** which carries blood to the left forelimb and left portion of the upper body. Locate the other axillary artery, the continuation of the left subclavian artery.

The aorta continues caudally along the dorsal body wall and passes through the diaphragm into the abdominal cavity. At this point it is commonly called the **dorsal aorta**. You will have to move the lobes of the left lung toward the center of the pig to view this. (DO NOT remove the lungs yet!)

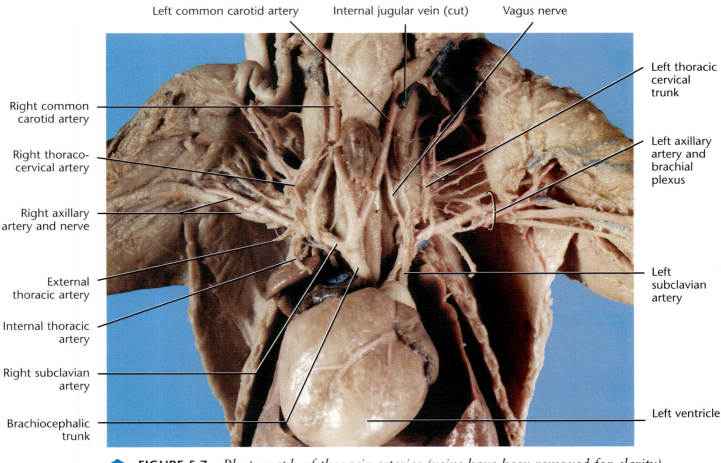

FIGURE 5.7   *Photograph of thoracic arteries (veins have been removed for clarity).*

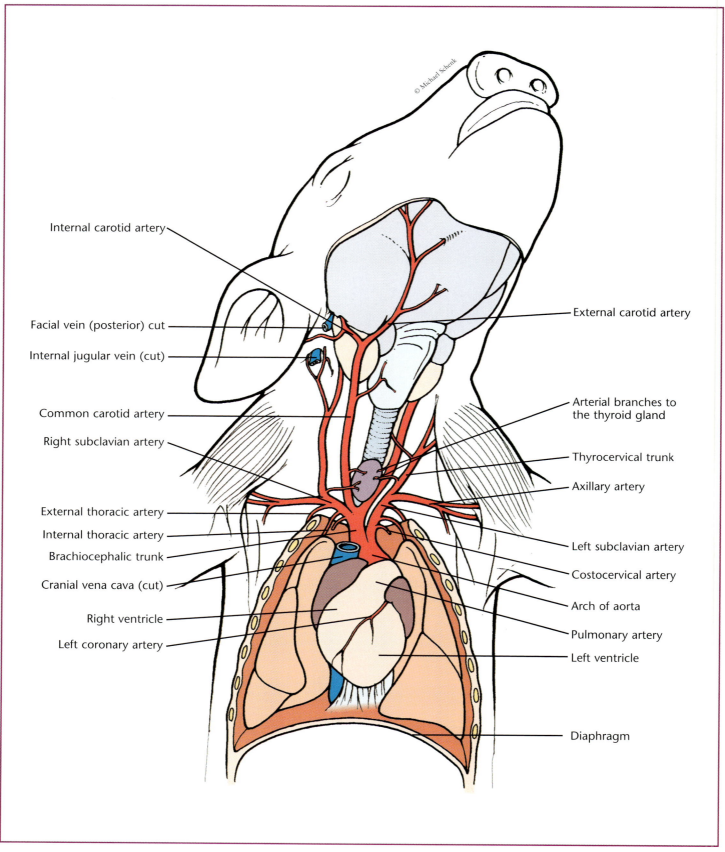

Internal carotid artery

External carotid artery

Facial vein (posterior) cut

Internal jugular vein (cut)

Arterial branches to the thyroid gland

Common carotid artery

Thyrocervical trunk

Right subclavian artery

Axillary artery

External thoracic artery

Internal thoracic artery

Left subclavian artery

Brachiocephalic trunk

Costocervical artery

Cranial vena cava (cut)

Arch of aorta

Right ventricle

Pulmonary artery

Left coronary artery

Left ventricle

Diaphragm

© Michael Schenk

FIGURE 5.8 *Illustration of thoracic arteries (veins have been omitted for clarity).*

After you have identified the major blood vessels in your pig, carefully remove the heart by cutting the (1) pulmonary arteries (near the ductus arteriosus), (2) aorta, (3) cranial and caudal vena cavae and (4) pulmonary veins. Place the heart in a dissecting pan and make a longitudinal cut along the frontal plane of the heart (dividing it into dorsal and ventral halves).

Notice that inside the chambers of the heart there are valves to prevent blood from flowing backwards (Figures 5.9 and 5.10). As blood enters the right atrium, it immediately flows into the right ventricle. Very little blood is actually *pumped* by the right atrium into the right ventricle. At the juncture of the right atrium and right ventricle is the **tricuspid valve**. As the right ventricle contracts and pushes blood out to the lungs, the blood is forced back up against the tricuspid valve, closing the leaflets of the valve and preventing retrograde flow into the right atrium. Upon entering the pulmonary trunk, the blood also passes through the **pulmonary semilunar valve**, which prevents backflow into the right ventricle. The blood returns from the lungs into the left atrium via the pulmonary veins and then flows into the left ventricle through the **bicuspid** (or **mitral**) valve. Blood leaving the left ventricle into the aorta is regulated by the **aortic semilunar valve**. The valves of the heart are prevented from being pushed too far backward (a condition known as "prolapse") by small stringlike attachments called **chordae tendineae**. You may be able to see these structures on your frontal section of the heart.

Abdominal Cavity

Hepatic Portal System

Follow the **caudal vena cava** from the heart through the diaphragm and liver toward the stomach. Notice how it passes directly through the diaphragm and through the center of the lobes of the liver. Between the liver and stomach there is a unique system of veins called the **hepatic portal system** (Figures 5.11 and 5.12). Portal systems in general are found in many different parts of the body in mammals. They serve important functions in rerouting blood to special organs before allowing it to pass along to the rest of the body. The difference in circulation between a normal circulatory pathway and a portal system is depicted below.

NORMAL CIRCULATORY PATHWAY:

artery ➡ capillary bed ➡ vein

PORTAL SYSTEM:

artery ➡ capillary bed ➡ portal vein ➡ capillary bed ➡ vein

In the case of the hepatic portal system, blood flows from the capillary beds of the jejunum, spleen and stomach into the **hepatic portal vein** and then into the capillary bed of the liver, before entering the caudal vena cava. This extra step allows blood from the stomach and intestine that contains large amounts of

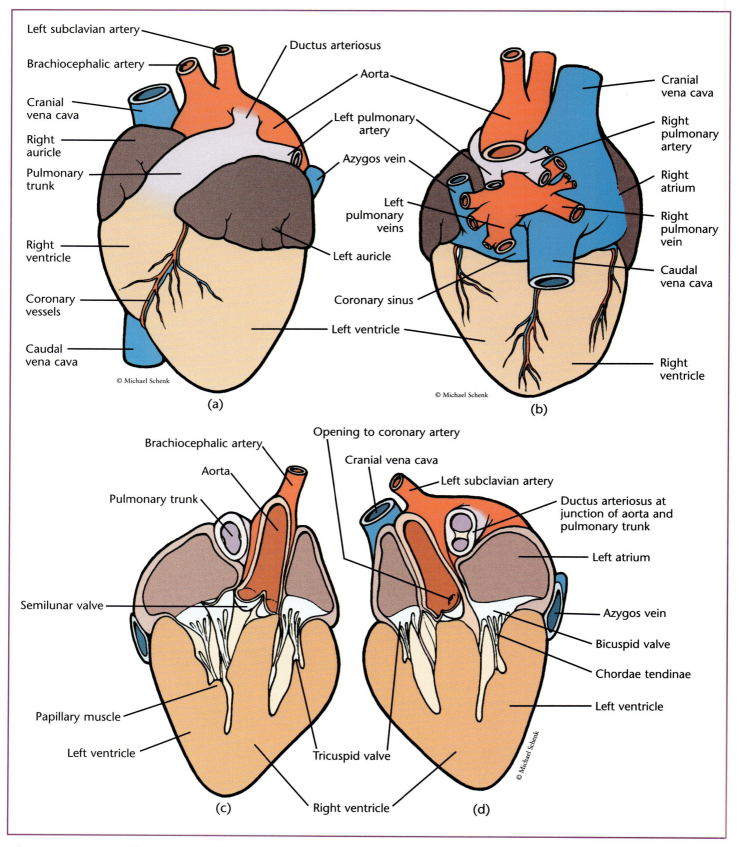

(a)

(b)

(c)

(d)

FIGURE 5.9 *Illustrations of heart depicting (a) ventrolateral view, (b) dorsal view and (c and d) interior views through frontal plane of heart.*

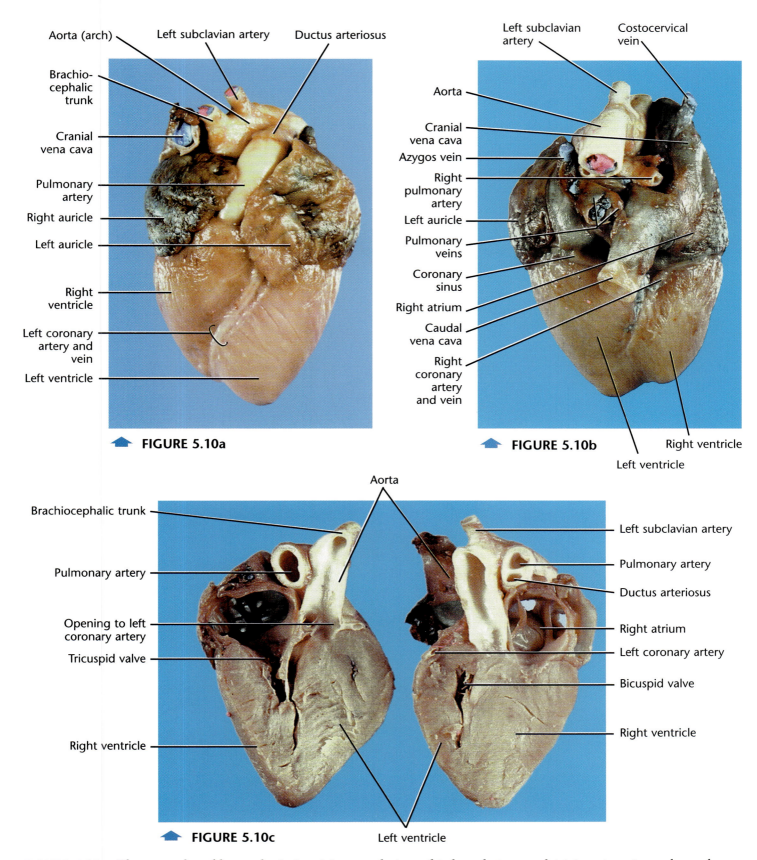

FIGURE 5.10a

FIGURE 5.10b

FIGURE 5.10c

FIGURE 5.10 *Photographs of heart depicting (a) ventral view, (b) dorsal view and (c) interior views through frontal plane of heart.*

sugars and possibly toxins and undesirable compounds to be filtered by the liver before the blood is sent to the rest of the body. This permits the liver to store the sugar (as glycogen) and release it into the blood as needed, thus maintaining a nearly constant level of glucose in the blood. Secondly, any deleterious compounds (like alcohol!) are removed from the blood before they reach other organs in the body and possibly harm them. Sometimes, however, the intake of these toxins exceeds the rate at which the liver can filter them from the blood, and the excess passes through the bloodstream to the other organs (like the brain).

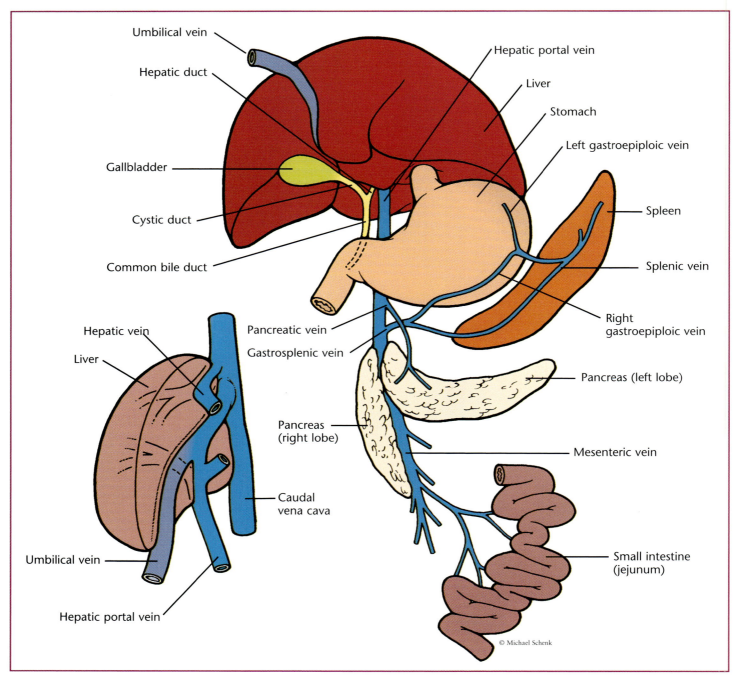

FIGURE 5.11   *Illustration of hepatic portal system depicting associated veins and organs.*

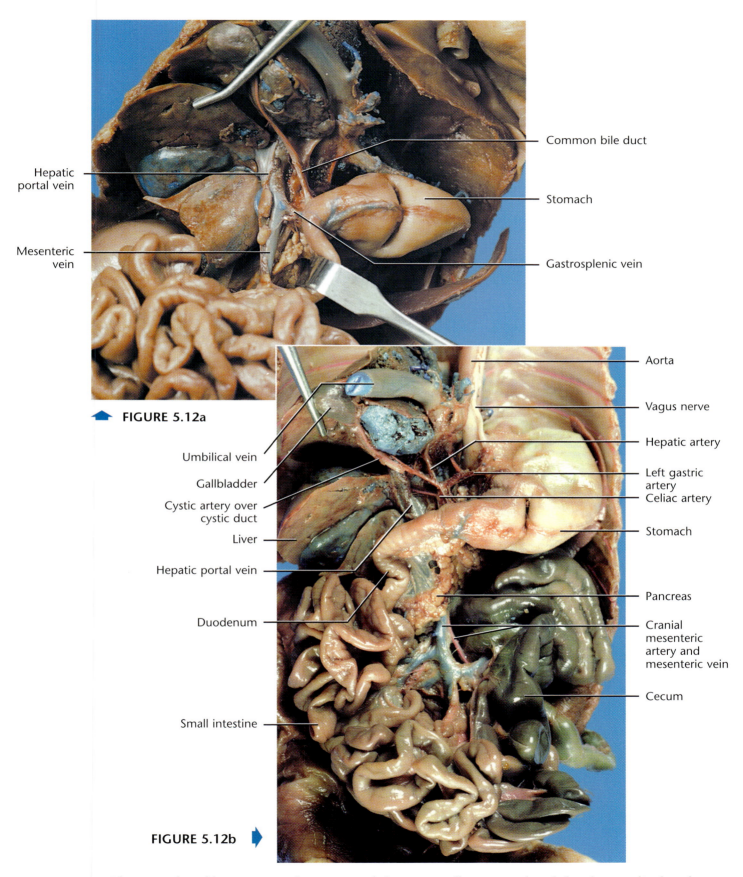

Hepatic portal vein

Mesenteric vein

Common bile duct

Stomach

Gastrosplenic vein

FIGURE 5.12a

Umbilical vein

Gallbladder

Cystic artery over cystic duct

Liver

Hepatic portal vein

Duodenum

Small intestine

Aorta

Vagus nerve

Hepatic artery

Left gastric artery

Celiac artery

Stomach

Pancreas

Cranial mesenteric artery and mesenteric vein

Cecum

FIGURE 5.12b

FIGURE 5.12   *Photographs of hepatic portal system with liver partially removed and duodenum displaced.*

Arteries and Veins of the Abdominal Region

As the aorta passes caudally through the abdominal region it makes several more branches (Figures 5.13–5.17). Locate the **celiac artery**, a small branch from the aorta to the stomach, pancreas and spleen. The **spleen** is a very important organ in the circulatory system. Its role is to store and release red blood cells into the bloodstream and recycle old red blood cells (Figure 5.18). Next find the **cranial mesenteric artery** (Figure 5.14a), which has branches that supply

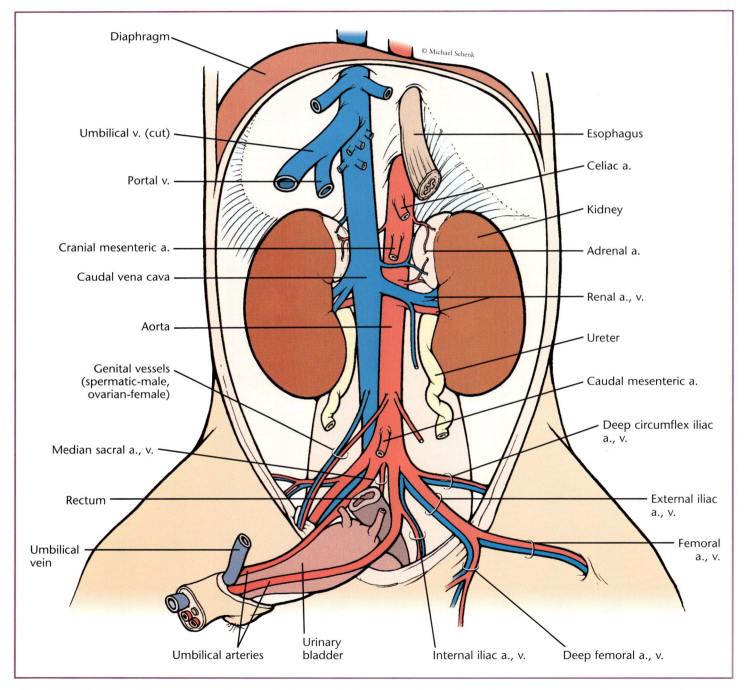

© Michael Schenk

FIGURE 5.13 *Illustration of ventral view depicting arterial supply and venous return of organs in the abdominal cavity and the lower extremities.*

the jejunum, ileum and colon. Embedded in the intestinal mesentery are the arterial arcades (Figure 5.14b), numerous branches of the mesenteric artery that provide nutrients and oxygen to the tissues of the intestinal tract. Further caudally, two short branches of the dorsal aorta lead into the kidneys. These are the **renal arteries**. Lying next to the renal arteries, the thinner-walled **renal veins** are present. These vessels collect filtered blood from the kidneys. The **caudal mesenteric artery** is a single vessel that supplies blood to the colon and rectum (Figure 5.15a). Caudal to this vessel, the **genital arteries** (Figure 5.15b) are

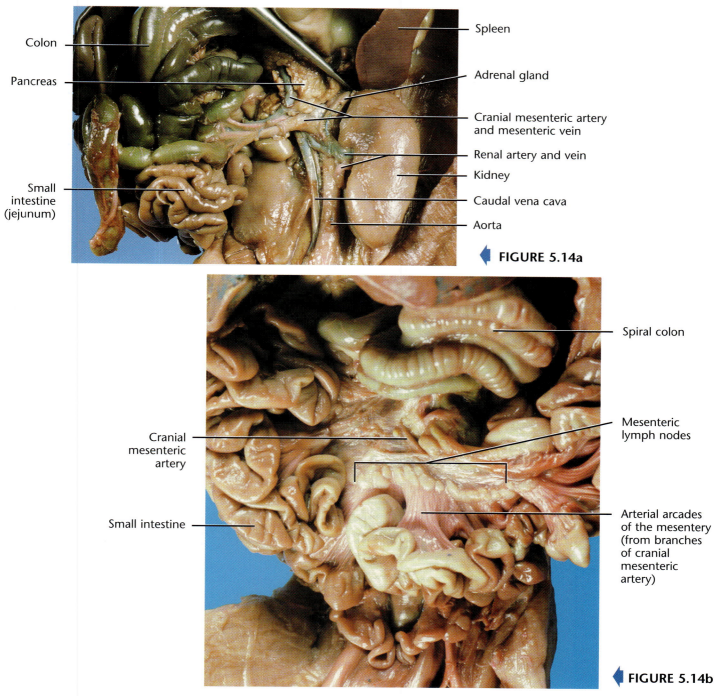

FIGURE 5.14a

FIGURE 5.14b

FIGURE 5.14   *Photographs of (a) cranial mesenteric artery, (b) arterial arcades of the cranial mesenteric artery, and (c) cranial mesenteric, celiac, and caudal mesenteric arteries, and their branches.*

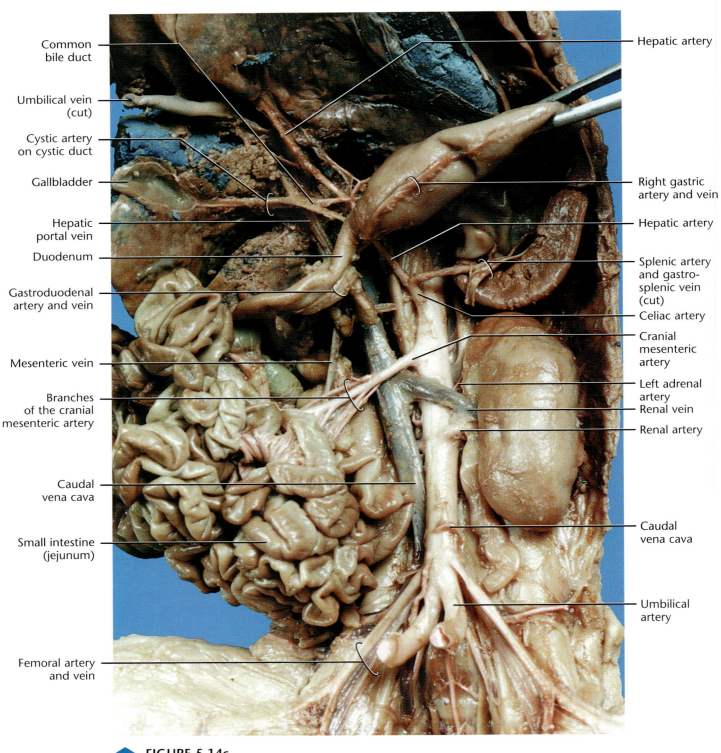

Common bile duct

Umbilical vein (cut)

Cystic artery on cystic duct

Gallbladder

Hepatic portal vein

Duodenum

Gastroduodenal artery and vein

Mesenteric vein

Branches of the cranial mesenteric artery

Caudal vena cava

Small intestine (jejunum)

Femoral artery and vein

Hepatic artery

Right gastric artery and vein

Hepatic artery

Splenic artery and gastro-splenic vein (cut)

Celiac artery

Cranial mesenteric artery

Left adrenal artery

Renal vein

Renal artery

Caudal vena cava

Umbilical artery

FIGURE 5.14c

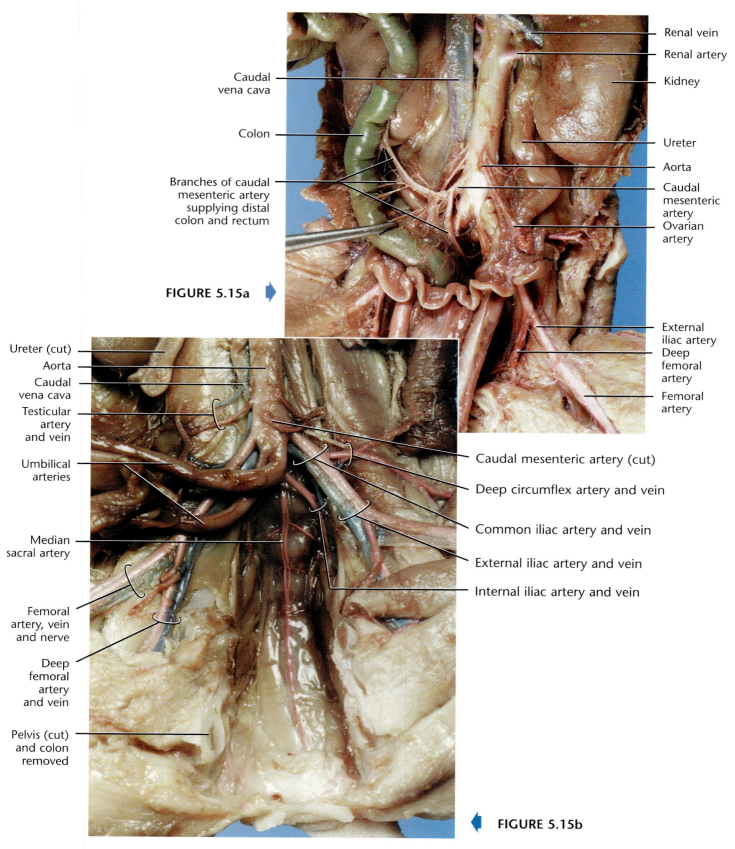

Renal vein

Renal artery

Kidney

Caudal vena cava

Colon

Ureter

Aorta

Branches of caudal mesenteric artery supplying distal colon and rectum

Caudal mesenteric artery

Ovarian artery

FIGURE 5.15a

External iliac artery

Deep femoral artery

Femoral artery

Ureter (cut)

Aorta

Caudal vena cava

Testicular artery and vein

Umbilical arteries

Caudal mesenteric artery (cut)

Deep circumflex artery and vein

Common iliac artery and vein

Median sacral artery

External iliac artery and vein

Internal iliac artery and vein

Femoral artery, vein and nerve

Deep femoral artery and vein

Pelvis (cut) and colon removed

FIGURE 5.15b

FIGURE 5.15 *Photographs of (a) caudal mesenteric artery and (b) genital, iliac, and femoral arteries (rectum cut and digestive tract removed for clarity).*

visible branching off the dorsal aorta and leading to the ovaries or testes, depending upon the sex of the pig. The **genital veins** run alongside the corresponding arteries but lead back into the caudal vena cava. If you follow the dorsal aorta caudally to the point where it branches into each hindlimb, you should be able to identify several more arteries and veins. Locate the **external iliac arteries** and **external iliac veins** leading into the upper thigh of each hindlimb. These vessels supply and receive blood from the legs. Further along, as they pass from the abdomen into the hindlimb, they become the **femoral arteries** and **femoral veins** and then branch into the **deep femoral arteries** and **deep femoral veins**. Finally, locate the **internal iliac arteries** and **internal iliac veins**, lying dorsal to the colon.

Umbilical Cord

> *In the abdominal region, locate the umbilical cord. You may need to make a fresh cut through the severed end of the cord to examine the internal structures in cross section.*

In placental mammals, the unborn young are attached to the placenta of the mother via the **umbilical cord** (Figure 5.3c). This cord serves as the lifeline for the fetus, transporting nutrients and oxygen to the growing fetus, and providing a channel for carbon dioxide and excess metabolic wastes to be eliminated from the fetus. Identify the single **umbilical vein**. Unlike most veins, this vessel actually carries oxygen- and nutrient-rich blood to the fetus from the fetal side of the placenta. The maternal blood is restricted to the maternal side of the placenta and never mixes with fetal blood under normal circumstances. Exchange

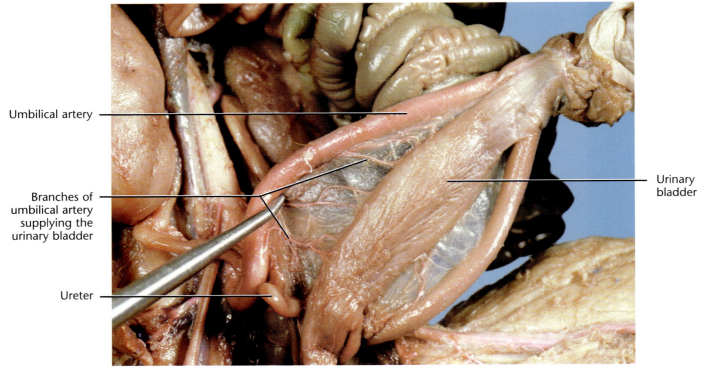

Umbilical artery

Branches of umbilical artery supplying the urinary bladder

Ureter

Urinary bladder

FIGURE 5.16   *Close-up of umbilical arteries leading into the umbilical cord.*

of gases, nutrients and wastes is accomplished by diffusion across the placental barrier. You should see two smaller **umbilical arteries**. These carry blood from the fetus to the placenta. The **allantoic duct** is important for eliminating metabolic waste (e.g., urine) from the fetal pig and channeling it to the allantois (a reservoir for toxic metabolic wastes in the uterus of the mother).

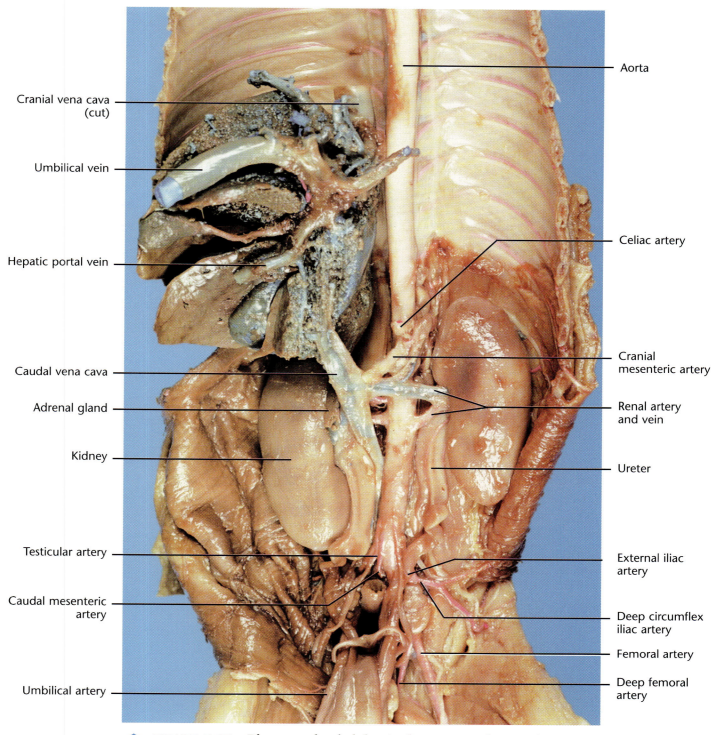

FIGURE 5.17 *Photograph of abdominal arteries and veins (liver partially removed and digestive tract removed).*

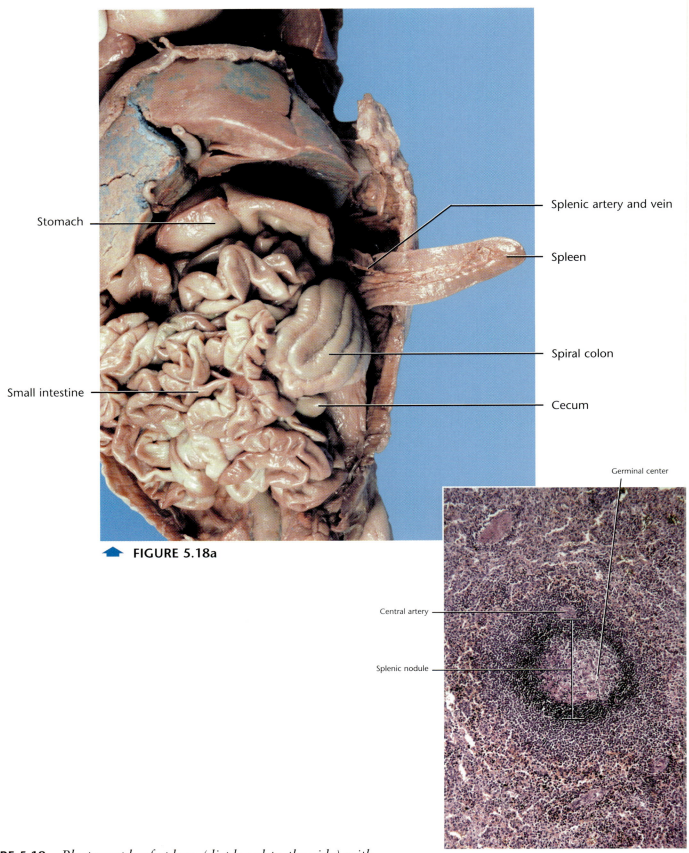

Stomach

Small intestine

Splenic artery and vein

Spleen

Spiral colon

Cecum

▲ **FIGURE 5.18a**

Germinal center

Central artery

Splenic nodule

FIGURE 5.18 *Photograph of spleen (displaced to the side) with accompanying histology photograph.*

▲ **FIGURE 5.18b** *Spleen.*   *250X*

RESPIRATORY SYSTEM

Laboratory Objectives

After completing this chapter, you should be able to:

1. Identify the major respiratory structures of the fetal pig.
2. Discuss the function of all indicated structures.
3. Discuss the flow of oxygen and carbon dioxide through the respiratory system.
4. Identify the microanatomy of respiratory tissue.

The respiratory system of mammals is responsible for bringing a fresh supply of oxygen to the bloodstream and carrying off excess carbon dioxide. The anatomy of the respiratory tract is designed to humidify and warm the air while filtering out dust particles and germs. The lining of the nasal epithelium is covered with fine hairs that capture these foreign particles and prevent them from passing into the lungs where they may infect the body. Similarly, as air is exhaled it is cooled and dried, thus reducing the amount of heat and moisture that terrestrial animals lose through respiration.

> *Using scissors, extend the midline incision (made earlier) by cutting cranially along the ventral midline of your pig from the top of the rib cage toward the chin. Be careful, many glands and organs lie just under the skin and may be damaged if you cut too deeply. Carefully tease away the surrounding tissue to expose the trachea.*

The Thoracic Cavity

The **trachea** is a long tube reinforced with cartilaginous rings to prevent collapse as the pig inhales (Figure 6.1). This tube leads from the nasopharynx (identified earlier) through the larynx ("voice box") and into the lungs. Locate the **larynx**. It should appear as an enlarged, oval-shaped protrusion toward the cranial end of the trachea. The larynx allows mammals to have a vast repertoire of vocalizations ranging from ultrasonic squeaks and chirps (in bats) and guttural barks or grunts (in dogs and pigs) to the highly complex sounds of

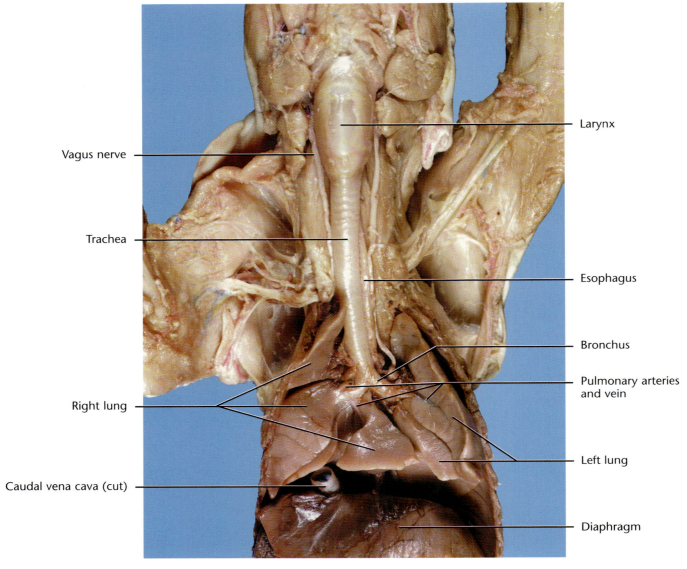

Vagus nerve

Trachea

Right lung

Caudal vena cava (cut)

Larynx

Esophagus

Bronchus

Pulmonary arteries and vein

Left lung

Diaphragm

FIGURE 6.1   *Photograph of respiratory system of pig (with heart removed).*

human speech. The pitch of these vocalizations is controlled by muscles attached to the larynx which contract and relax, altering the shape of the voice box, thus changing the sounds that it produces. Follow the trachea caudally toward the lungs. Notice where it splits into three **bronchi** (Figure 6.2–6.3). These lead into the **left and right lung**. Notice that the right lung has four lobes, while the left lung has three. (In humans, the right lung has three lobes, while the left has two.) Identify the **right cranial lobe**, **right medial lobe**, **right caudal lobe** and **accessory lobe**. On the left side of the pig, the **left cranial lobe** and **left medial lobe** are actually attached to form a common lobe, called the left apical lobe. Also identify the **left caudal lobe**. Underneath the lungs you should be able to see a thin muscular sheet of tissue, the **diaphragm**. This structure (unique to mammals) allows the thoracic cavity to expand and compress, drawing in fresh air with each expansion (as the diaphragm contracts) and expelling stale air with each compression (as the diaphragm relaxes).

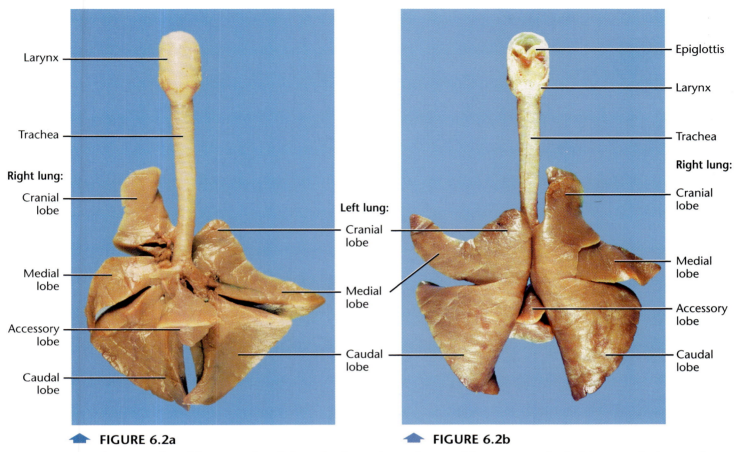

Larynx

Trachea

Right lung:

Cranial
lobe

Medial
lobe

Accessory
lobe

Caudal
lobe

Left lung:

Cranial
lobe

Medial
lobe

Caudal
lobe

Epiglottis

Larynx

Trachea

Right lung:

Cranial
lobe

Medial
lobe

Accessory
lobe

Caudal
lobe

▲ **FIGURE 6.2a**   ▲ **FIGURE 6.2b**

FIGURE 6.2   *Photographs of lungs isolated from body with trachea and larynx attached, (a) ventral view and (b) dorsal view.*

Inside the lungs, the bronchi are further divided into several branches called **bronchioles** (Figure 6.3). These bronchioles branch into smaller and smaller tubules, eventually terminating in open sacs called **alveoli**. These alveoli are comprised of thin epithelial tissue and are surrounded by capillary networks. It is here, in the adult, where oxygen is picked up by the bloodstream and carbon dioxide is released back into the lungs to be expelled from the body with each exhalation.

The Oral Cavity

As the pig inhales, air is taken in through the external **nares** and passes through the **nasopharynx**. At this point, the **glottis** is "open," with the **epiglottis** permitting air flow through the **larynx** into the **trachea** (Figure 6.4A). However, when the pig swallows, food passes through the oral cavity (on the ventral side of the hard and soft palate) and is prevented from entering the respiratory tract by the action of the epiglottis closing to cover the entrance through the glottis (Figure 6.4B). In the evolution of vertebrates, the advent of the complete secondary palate (the continuous hard and soft palate) was a major advancement. Animals could now eat with no loss in respiratory ability, since the complete secondary palate effectively keeps the food passageway and airway separated.

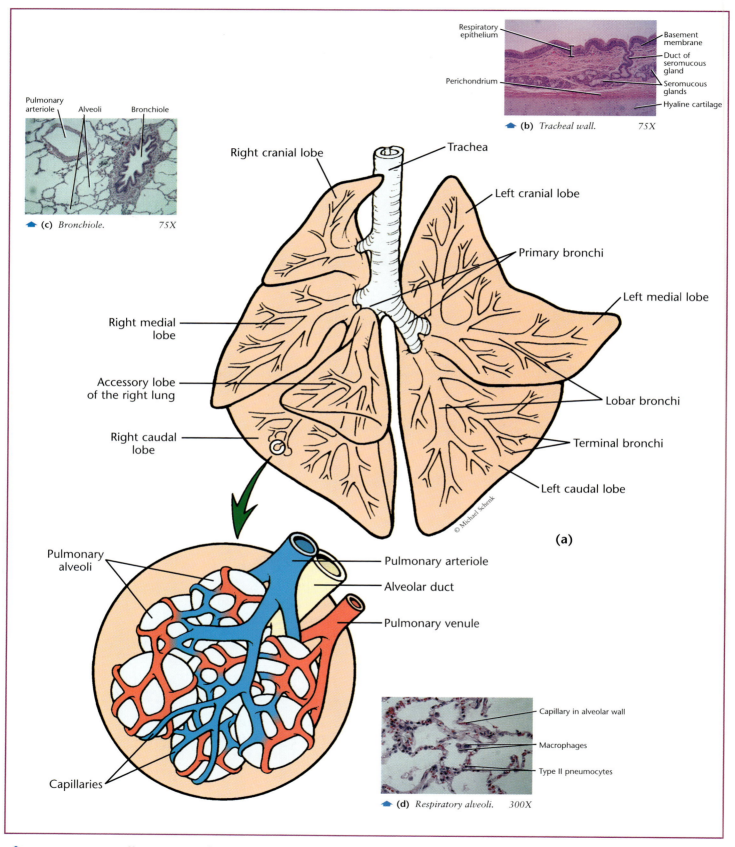

Respiratory epithelium

Basement membrane

Duct of seromucous gland

Perichondrium

Seromucous glands

Hyaline cartilage

(b) *Tracheal wall.*   *75X*

Pulmonary arteriole

Alveoli

Bronchiole

(c) *Bronchiole.*   *75X*

Right cranial lobe

Trachea

Left cranial lobe

Primary bronchi

Left medial lobe

Right medial lobe

Accessory lobe of the right lung

Lobar bronchi

Right caudal lobe

Terminal bronchi

Left caudal lobe

© Michael Schenk

(a)

Pulmonary alveoli

Pulmonary arteriole

Alveolar duct

Pulmonary venule

Capillaries

Capillary in alveolar wall

Macrophages

Type II pneumocytes

(d) *Respiratory alveoli.*   *300X*

FIGURE 6.3   *Illustration of (a) lungs showing alveolar sacs; histology photographs of (b) tracheal wall, (c) bronchiole and (d) alveoli.*

Many reptiles must pause while eating, take a few deep breaths, and then resume swallowing their food. Overcoming this constraint was one of many characteristics which contributed to mammals' ability to maintain a high metabolic rate and be endothermic. Endothermy requires a high metabolism which requires an animal to consume large amounts of nutrients.

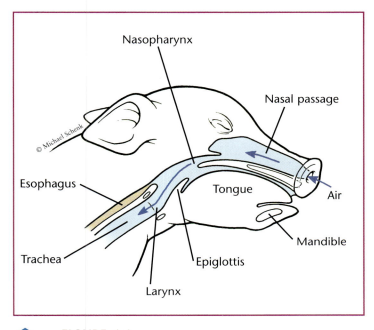

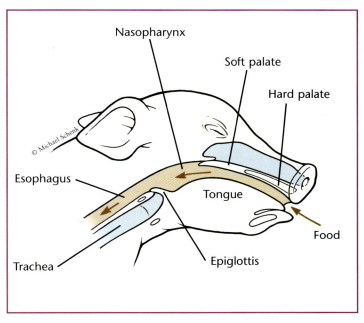

FIGURE 6.4a

FIGURE 6.4b

Illustrations depicting mechanisms of (a) breathing and (b) swallowing in the pig.

CHAPTER 7

REPRODUCTIVE AND EXCRETORY SYSTEMS

Laboratory Objectives

After completing this chapter, you should be able to:

1. Identify the major reproductive structures of both male and female pigs.
2. Identify the major excretory structures of the fetal pig.
3. Discuss the function of all indicated structures.
4. Identify the microanatomy of the excretory and reproductive organs.

The reproductive system is responsible for producing **gametes** that will ultimately fuse with the corresponding gamete of the opposite sex. For this reason, the gametes are produced through a special kind of cell division (meiosis) that reduces their chromosome number by half. Fertilization restores the chromosome number of the embryo to the normal number. Because the continual existence of a species depends solely on its individuals' success at passing genes on to the next generation, reproductive systems have evolved highly complex features that increase the chances of fertilization and successful embryonic development.

Extend the incision made earlier along the ventral midline of the abdominal region from the umbilical cord caudally toward the anus, using scissors. Cut just slightly to one side of the midline to avoid cutting through the urethra and other organs. You will need to apply a bit of pressure to cut through the pelvis. If you have a male pig, continue on with the next section. If you have a female pig, skip ahead to the section entitled "Female Reproductive System." However, regardless of the sex of your pig, you are expected to be familiar with the structures of each sex, so work closely with another group that has a pig of the opposite sex.

Male Reproductive System

If your male pig is sufficiently mature, **scrotal sacs** may be present (Figures 7.1 and 7.2). The **testes** will be found along the dorsal side of the abdominal cavity somewhere between the base of the kidneys and the scrotal sacs (depending on how far they have migrated). They are small bean-shaped structures. Cupped around the side of each testis is a highly-coiled system of tubules known as the **epididymis**. Sperm are produced in the **seminiferous tubules** of the testes and are stored in the epididymis. The sperm travel through the **ductus deferens** toward the **urethra**. Careful dissection will reveal several accessory glands along this route. At the juncture of the ductus deferens and the urethra, identify the **seminal vesicles**. The **bulbourethral glands** lie caudally, on either side of the urethra. Together, these glands contribute fluid to the sperm, over 60% of the total volume of the semen. This fluid is thick and contains mucus (to prevent the

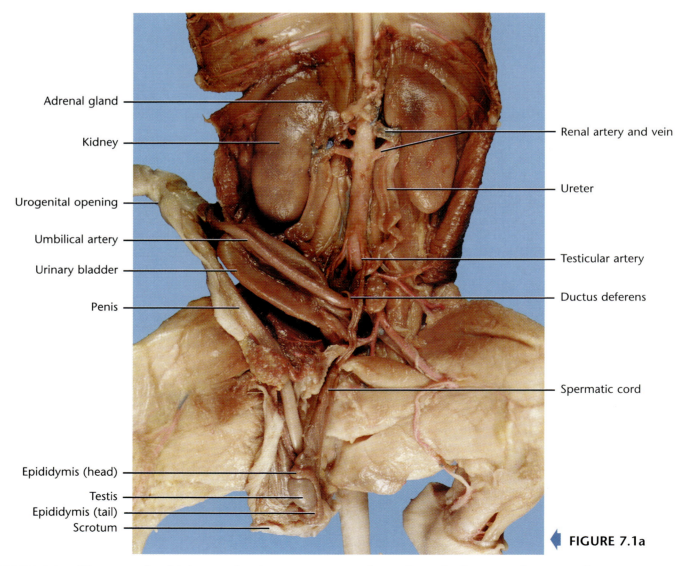

Adrenal gland

Kidney

Urogenital opening

Umbilical artery

Urinary bladder

Penis

Epididymis (head)

Testis

Epididymis (tail)

Scrotum

Renal artery and vein

Ureter

Testicular artery

Ductus deferens

Spermatic cord

FIGURE 7.1a

FIGURE 7.1   *Photograph of (a) reproductive structures in the male with close-ups depicting (b) testis in scrotal sac and (c) seminal vesicles and bulbourethral gland (digestive tract removed for clarity).*

(continued on next page)

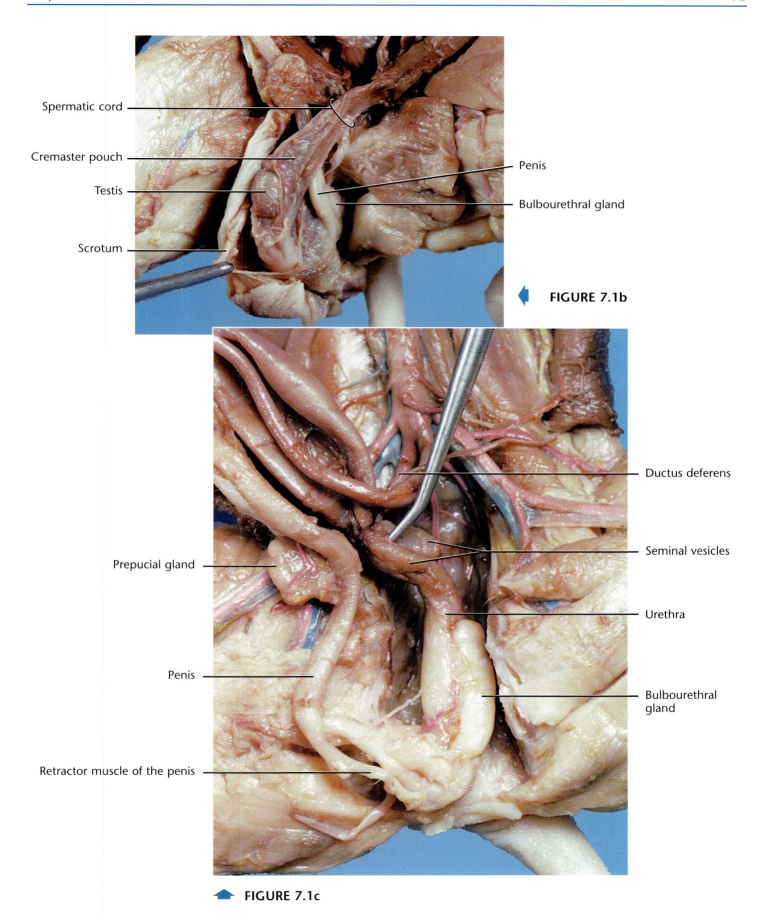

Spermatic cord

Cremaster pouch

Testis

Scrotum

Penis

Bulbourethral gland

◀ **FIGURE 7.1b**

Ductus deferens

Prepucial gland

Seminal vesicles

Urethra

Penis

Bulbourethral gland

Retractor muscle of the penis

▲ **FIGURE 7.1c**

sperm from drying out) and large amounts of fructose (to provide energy for the sperm). In addition, the semen is highly basic to neutralize the acidic environment of the vagina and increase the chances of survival for the sperm. The semen passes through the **urethra** and into the **penis** from which it will be ejaculated. Notice that the penis of the fetal pig does not protrude from the body yet. To locate it, you will need to carefully tease the penis away from the tissue along the ventral midline of the body.

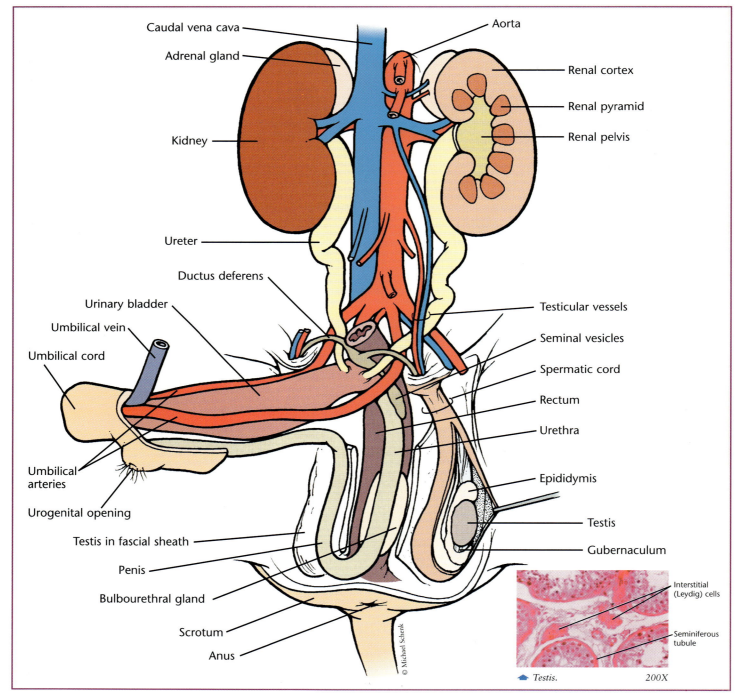

FIGURE 7.2   *Schematic illustration of male genitalia with histology photograph of testis; digestive tract omitted for clarity.*

Female Reproductive System

The paired female gonads are called **ovaries** (Figures 7.3, 7.4, and 7.5). They are located in the abdominal region caudal to the kidneys and can be identified by their small, round appearance. Attached to each ovary is a coiled **oviduct**. The oviduct receives the mature oocyte (egg) when it is released from the ovary

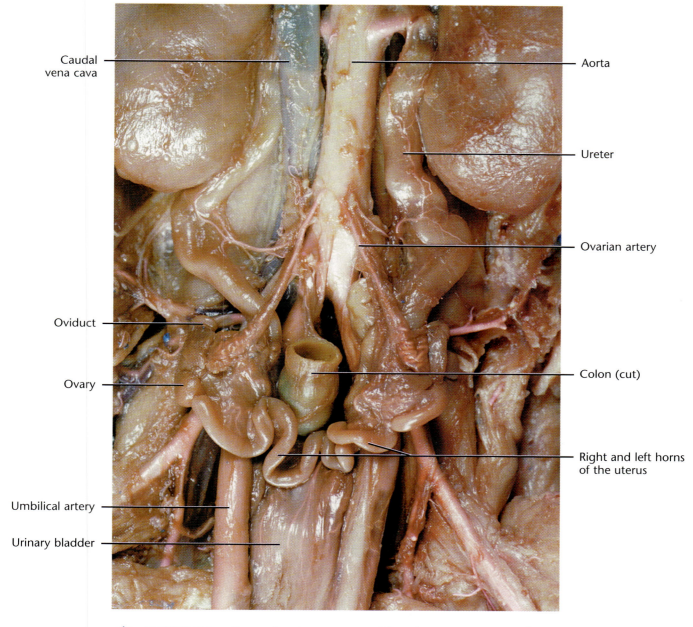

Caudal vena cava

Aorta

Ureter

Ovarian artery

Oviduct

Colon (cut)

Ovary

Right and left horns of the uterus

Umbilical artery

Urinary bladder

FIGURE 7.3 *Reproductive system of female; rectum cut and digestive system removed for clarity.*

at the time of ovulation. The epithelial linings of the oviducts are ciliated and propel the eggs along the lengths of the oviducts toward the horns of the uterus. Fertilization typically occurs in the oviducts, but implantation of the embryos occurs in the uterus. In pigs, the uterus is divided into two **uterine horns** and a **body of the uterus**. Due to their large litter sizes (up to 14 off-spring), female pigs require a large area in the uterus for young to develop. The body of the uterus extends caudally to the **cervix**. Locate the juncture of the cervix and the **vagina**. The cervix is a constriction of semi-cartilaginous tissue, while the vagina extends caudally from this constriction. The vagina is joined by the urethra and the two open into a common chamber called the **urogenital**

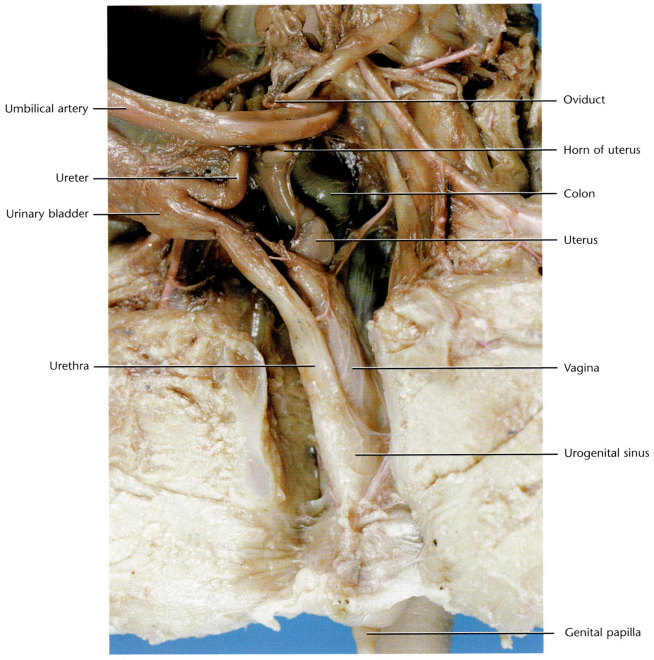

Umbilical artery

Ureter

Urinary bladder

Urethra

Oviduct

Horn of uterus

Colon

Uterus

Vagina

Urogenital sinus

Genital papilla

FIGURE 7.4 *Reproductive system of female; rectum cut and digestive system removed for clarity.*

sinus (since it handles products of both the urinary and reproductive systems). The urogenital sinus opens to the outside of the body through the **urogenital opening**. The **genital papilla** will be visible as a small projection on the caudoventral surface of the abdominal cavity covering the urogenital opening. In humans this structure develops into the clitoris and plays a role in sexual sensation.

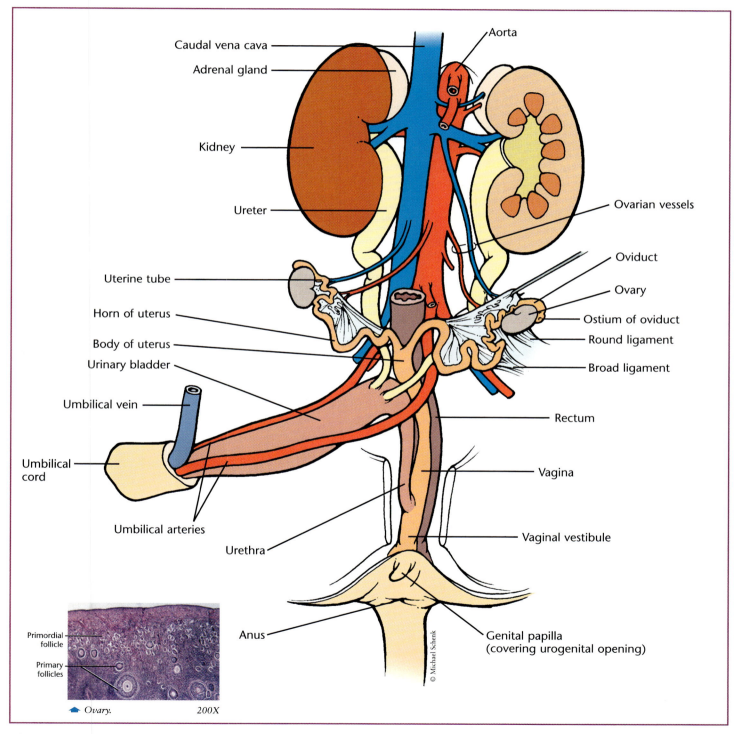

FIGURE 7.5 *Illustration of female reproductive system with histology photograph of ovary.*

TABLE 7.1 *Male and female reproductive organs and their functions. Corresponding homologous structures in the two sexes are placed in the same row. Structures denoted with an asterisk (*) are accessory glands.*

Male Structure	Function	Female Structure	Function
Testis	Produces sperm.	Ovary	Produces eggs.
Epididymis	Stores sperm.	Oviduct	Receives egg at ovulation; site of fertilization.
Ductus deferens	Transports sperm to urethra.	Uterine horns (and body)	Site of implantation and embryonic development.
Urethra	Receives seminal secretions from accessory glands.	Urethra	Drains excretory products from urinary bladder (no reproductive function).
Penis	Deposits semen in female reproductive tract.	Genital papilla	Develops into the clitoris (in humans).
Seminal vesicles*	Contribute seminal fluid containing nutrients for sperm, and hormones to stimulate uterine contractions.		
Bulbourethral glands*	Contribute seminal fluid that may aid in neutralization of acidity of vagina.		

Excretory System

> *The excretory systems of both the male and female pig are identical, so no special arrangements are necessary for viewing a pig of the opposite sex. Using a teasing needle, carefully dissect away the membranous tissue surrounding one of the kidneys. Take care not to destroy the adrenal gland which sits along the medial margin of the kidney. Clean the area around the kidney to expose the ureter passing from the medial margin of the kidney caudally toward the urinary bladder.*

The excretory system is responsible for eliminating the metabolic wastes that the body produces from cellular respiration. Remember, this is an entirely different process from that which expels undigested foodstuffs through the anus! Excretion and egestion are different processes, handled by completely different systems in the pig (and most other animals).

The **kidneys** are large, bean-shaped organs that lie along the dorsal surface of the abdominal cavity on either side of the spine (Figure 7.6). You have already seen the large **renal arteries** and **renal veins** that carry blood into and out of the kidneys. These organs filter blood from the circulatory system, removing the metabolic waste products produced in the tissues of the body during cellular respiration. Their major function is to concentrate these toxins and eliminate them from the body while conserving water, salts and other compounds that the body needs. In humans, the kidneys filter anywhere from 1100 to 2000 liters of blood each day! From this tremendous volume of blood only about 1.5 liters of urine is actually produced. The other 99.9% is reabsorbed into the bloodstream. The urine is concentrated in the kidneys and passes down the **ureter** toward the **urinary bladder**. Notice the unusual shape of the urinary bladder in the fetal pig. This is due to the nature of the fetus' reliance on its mother to remove metabolic wastes from its body. As soon as the fetus is born, the umbilical cord deteriorates and the fetus no longer eliminates wastes through the umbilical cord. Instead, the urine which is stored in the urinary bladder is eliminated through the **urethra** and out of the **urogenital opening**.

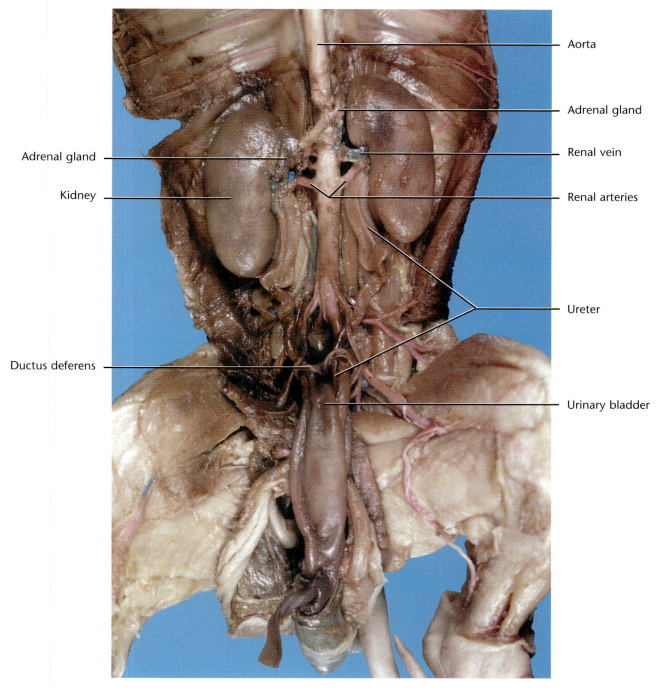

FIGURE 7.6   *Photograph of excretory system of fetal pig; digestive tract removed for clarity.*

> *Carefully remove one of the kidneys from your pig and make a longitudinal incision through the frontal plane cutting it into two equal halves (a dorsal and ventral half).*

The kidney is comprised of three major regions internally. The outer region, the **cortex**, the middle region, the **medulla**, and the inner-most region, the **renal pelvis**. Locate these regions on the frontal section of the kidney you bisected (Figure 7.7). Notice how the renal pelvis drains into the ureter. The renal pelvis collects the waste that is filtered from the blood. The functional unit of the mammalian kidney is called the **nephron**. The nephron is comprised of many substructures which together filter nitrogenous wastes from the blood while conserving valuable sugars, ions and water. Blood enters each nephron through an **afferent arteriole** that forms a capillary bed known as the **glomerulus**. Here, blood pressure forces water, urea, salts and other small soluble compounds from the blood into the epithelial lining of **Bowman's capsule**. Blood fluid that is not filtered out travels through an **efferent arteriole** to a capillary bed surrounding the convoluted tubules known as the **peritubular arteries**. Bowman's capsule receives the fluid and transports it along a series of **proximal convoluted tubules**, down the **loop of Henle** and through another series of **distal convoluted tubules**. During this stage of the filtration process, water, sodium chloride and potassium ions are reabsorbed into the bloodstream. This process produces a highly concentrated urine that passes into a **collecting duct**. Many nephrons converge at the collecting ducts which drain the kidneys. The urine passes from the kidney into the **ureter** and on to the **urinary bladder** for storage. Later, the urine will be eliminated from the pig through the **urethra**.

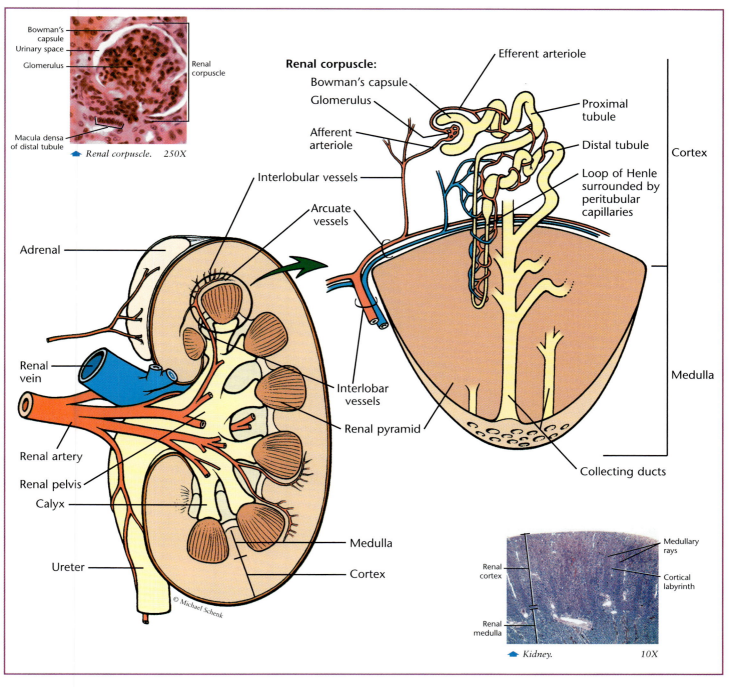

Renal corpuscle:
- Bowman's capsule
- Urinary space
- Glomerulus

Macula densa of distal tubule

Renal corpuscle

Renal corpuscle. 250X

Renal corpuscle:
- Bowman's capsule
- Glomerulus
- Afferent arteriole

Efferent arteriole

Proximal tubule

Distal tubule

Loop of Henle surrounded by peritubular capillaries

Cortex

Medulla

Interlobular vessels

Arcuate vessels

Adrenal

Renal vein

Renal artery

Renal pelvis

Calyx

Ureter

Interlobar vessels

Renal pyramid

Medulla

Cortex

Collecting ducts

© Michael Schenk

Renal cortex

Renal medulla

Medullary rays

Cortical labyrinth

Kidney. 10X

FIGURE 7.7 *Illustration of interior anatomy of kidney with nephron and histology photographs.*

TABLE 7.2 *Subunits of the mammalian kidney and urinary system and their functions.*

Organ/Structure	Function
Renal artery	Supplies blood to the kidney.
Renal vein	Transports filtered blood away from kidney to vena cava.
Afferent arteriole	Brings blood to each nephron to be filtered.
Efferent arteriole	Carries unfiltered portion of blood away from glomerulus to the capillary beds surrounding convoluted tubules and loop of Henle.
Glomerulus	Capillary bed that forces fluid containing salts, glucose, vitamins and nitrogenous wastes out of the bloodstream.
Bowman's capsule	Epithelial layer surrounding glomerulus that receives filtrate from the glomerulus.
Proximal convoluted tubules	Series of tubules that selectively remove sodium chloride, potassium, water and other nutrients from the nephron and return them to the bloodstream.
Peritubular arteries	Capillary bed surrounding convoluted tubules and loop of Henle.
Loop of Henle	Long extension of the nephron tubule that descends into the medulla of the kidney forming a concentration gradient which removes more water and sodium chloride, and produces a highly-concentrated urine.
Distal convoluted tubules	Series of tubules that selectively remove more water and sodium chloride, but absorb potassium.
Collecting ducts	Several nephrons converge on a single collecting duct, which further concentrates urine while passing it along to the ureter.
Ureter	Transports urine to the urinary bladder.
Urinary bladder	Stores urine.

NERVOUS SYSTEM

Laboratory Objectives

After completing this chapter, you should be able to:

1. Identify the major nerves of the fetal pig.

2. Describe the organization of the pig brain and spinal cord.

3. Identify the major structures in the sheep eye and describe their roles in vision.

The nervous system is devoted to receiving physical stimuli from the environment, converting it into electrical impulses, processing the information and effecting behavioral or physiological changes in response to the stimuli. The nervous system is divided into two main regions: the central nervous system (CNS), composed of the brain and spinal cord, and the peripheral nervous system (PNS), which includes the cranial nerves and spinal nerves emanating from the brain and spinal cord, respectively. The nerves of the PNS receive external stimuli (through sensory neurons) and produce motions in the muscles (through motor neurons). The brain and spinal cord are the sites of integration of the information picked up by the sensory neurons. These individual nerve cells are networked together to produce a highly complex, intricately organized system for communication and information transfer.

The Brain

Lay your pig on its ventral side and make a longitudinal incision with a scalpel through the skin covering the head starting at the base of the neck and continuing rostrally. Gently separate the skin from the skull using a blunt probe. Once the skull is completely exposed from the base of the neck to just in front of the eyes, begin shaving away at the top of the skull with a scalpel. When you penetrate the skull, begin "chipping" away pieces of the skull from this central opening, moving outward toward the perimeter of the brain case. The brain and spinal cord will be covered with a soft, clear series of membranes called the meninges which in life are filled with fluid to dampen vibrations and cushion the brain against jarring movements. You will need to remove the meninges with a blunt probe or teasing needle to reveal the brain. Be careful, the brain is very soft and delicate! Continue in this fashion until you have exposed the dorsal and lateral portions of the brain, brainstem and base of the spinal cord.

The **cerebrum** (the largest portion of the brain) is composed of **the left and right hemispheres** (Figures 8.1 – 8.3). Running between these two hemispheres is the **longitudinal fissure**. Internally the two hemispheres are joined together by the corpus callosum. The cerebrum is composed of several regions (or lobes): the **frontal region**, the **temporal region**, the **parietal region** and the **occipital region**. The cerebrum functions in the interpretation of sensory impulses and the coordination of voluntary movements. The parts of the brain responsible for higher

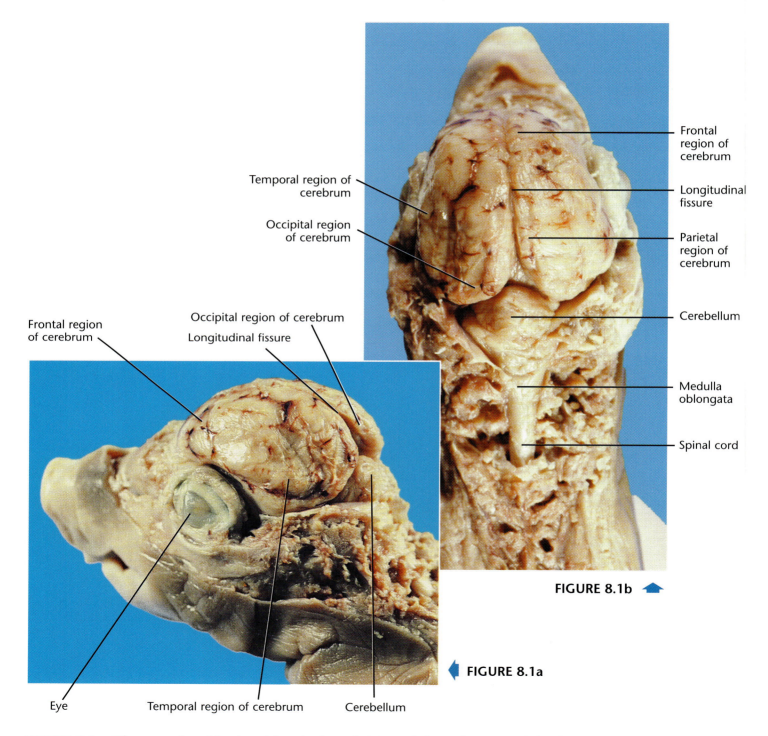

FIGURE 8.1b ⬆

FIGURE 8.1a ⬅

FIGURE 8.1   *Photographs of brain with spinal cord, (a) caudolateral view and (b) dorsal view.*

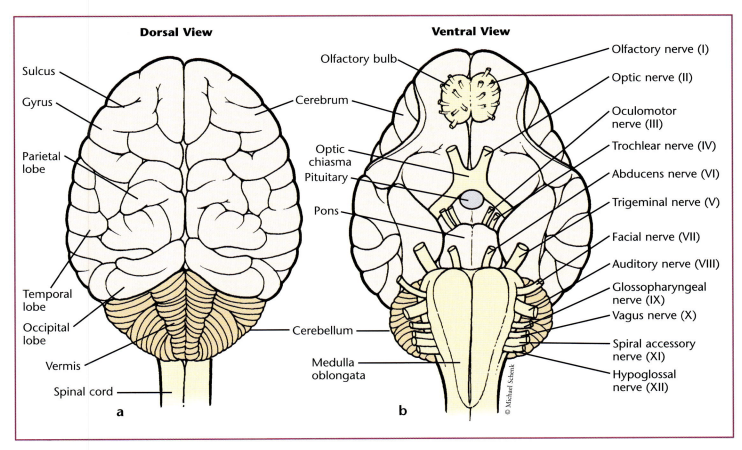

Dorsal View

Sulcus

Gyrus

Parietal lobe

Temporal lobe

Occipital lobe

Vermis

Spinal cord

a

Ventral View

Olfactory bulb

Cerebrum

Optic chiasma

Pituitary

Pons

Cerebellum

Medulla oblongata

b

© Michael Schenk

Olfactory nerve (I)

Optic nerve (II)

Oculomotor nerve (III)

Trochlear nerve (IV)

Abducens nerve (VI)

Trigeminal nerve (V)

Facial nerve (VII)

Auditory nerve (VIII)

Glossopharyngeal nerve (IX)

Vagus nerve (X)

Spiral accessory nerve (XI)

Hypoglossal nerve (XII)

FIGURE 8.2   *Illustrations of brain: dorsal view and ventral view showing the twelve cranial nerves.*

functions like memory and learning are located in the cerebrum. Caudal to the cerebrum is the smaller **cerebellum**. The cerebellum is primarily a reflex center for the integration of skeletal muscle movements. It is responsible for such activities as muscle coordination and balance. At the base of the cerebellum, locate the brainstem or **medulla oblongata**. This is the most caudal portion of the brain and leads into the **spinal cord**. The medulla oblongata is responsible for regulating many autonomic functions such as breathing, heart rate, digestion, sweating and vomiting.

Lateral View

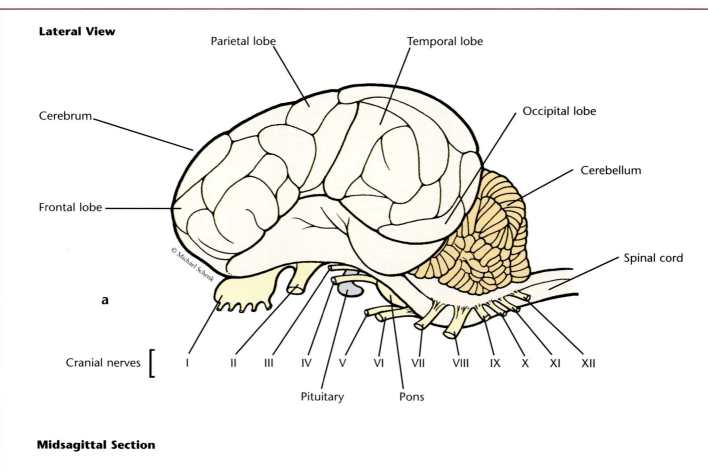

Midsagittal Section

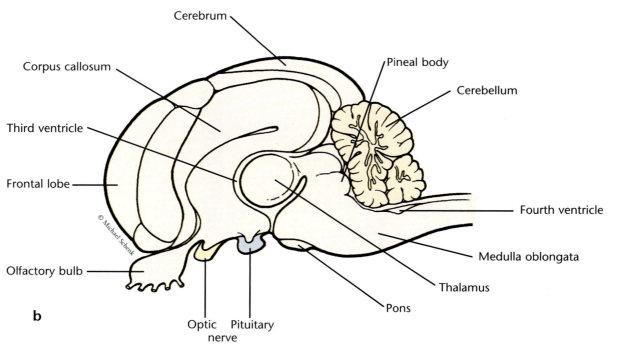

◄ **FIGURE 8.3** *Illustrations of brain: lateral view showing the twelve cranial nerves and midsagittal section.*

The Eye

The eye is a complex sensory organ equipped for receiving external stimuli (in the form of light waves) and converting this light energy into chemical information that can be integrated (to some extent by the eye, itself) and sent to the brain for interpretation. Since fetal pigs have eyes that are quite small, you will dissect the eye of a sheep (or cow) to acquire an appreciation for the intricate structures that play a role in the vision of mammals. Many of the anatomical features of the sheep eye are, in fact, quite similar to those of a human eye.

> *Begin by removing the fatty tissue that covers a large part of the eye. You will encounter bands of muscle tissue as well that may be cleaned away. Be careful not to puncture through the surface of the eye while preparing the external surface. With forceps, locate the optic nerve (on the opposite side of the eye from the clear cornea) and trim the fatty tissue away from it, being careful not to actually disturb the optic nerve.*

Many mammals, including pigs, possess a "third eyelid" that may still be present on your preserved eye. This eyelid, which is clear and remains invisible when closed, is known as the **nictitating membrane**. Humans lack a nictitating membrane; however a vestigial remnant of this structure is present. Identify the **cornea**—a tough, transparent layer that allows light to enter the eye while protecting the underlying structures (Figure 8.4). The cornea is composed of a special lamellar arrangement of cells that permits nearly perfect optical transparency. This clarity comes with a price however. The cells of the cornea must continuously pump out their interstitial fluid to maintain the proper structural arrangement necessary for clear vision. The **optic nerve** in the back of the eye is the site at which the axons of all the photoreceptors contained in the eye converge and send their information from the eye to the brain. Surrounding the remainder of the eye (exclusive of the cornea) and the optic nerve is a tough, white layer of tissue called the **sclera**. The sclera also helps protect the eye from damage.

> *Using scissors, carefully cut the eye in half through its frontal plane (giving you one half containing the cornea and one half with the sclera and the optic nerve stub). Place the eye in your dissecting tray with the cornea facing up and gently open it by lifting the front half.*

Inside the eye you should see the lens floating in a fluid-filled chamber known as the **vitreous chamber**. The fluid contained in this chamber is a mixture of water, called the **vitreous humor**, and fine transparent fibers suspended in the fluid. The **lens** is a fairly solid, biconvex structure composed of concentric sheets of clear cells (arranged much like the skin of an onion). While quite sturdy, the lens is flexible and capable of bending to focus the image on the **retina** at the back of the eyeball. Small muscles known as **ciliary bodies** attached to the lens accomplish this task. The rods and cones (photoreceptors) are imbedded in the retina. The back of the retina is covered with a reflective membrane called the **choroid layer** which enhances the amount of light that is

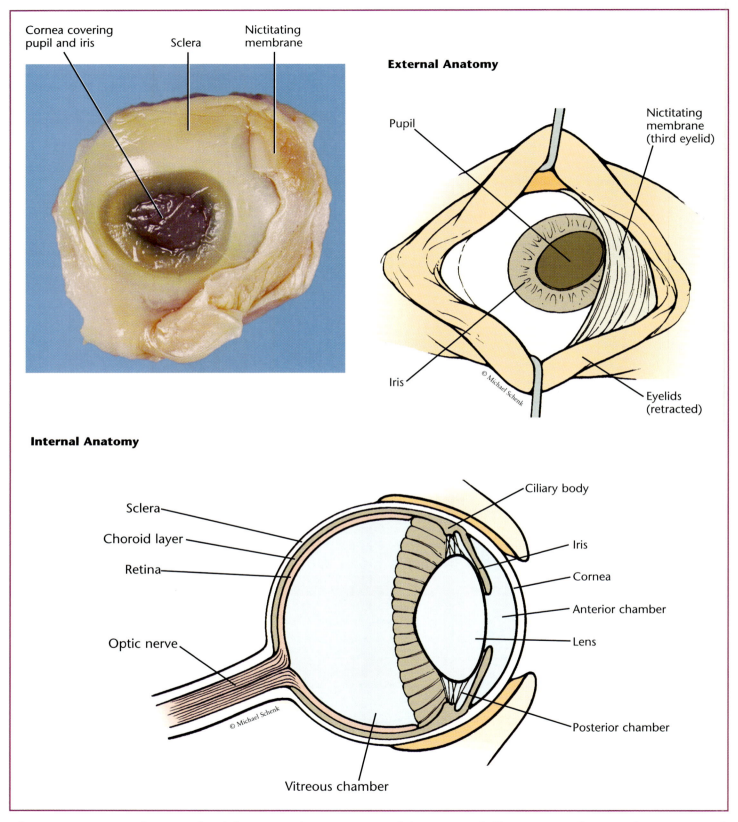

External Anatomy

Internal Anatomy

FIGURE 8.4   *Photograph of sheep eye showing external features and illustrations of external anatomy and internal anatomy of the eye.*

reflected back into the rods and cones of the eye. One of the most amazing features of the mammalian eye is the way in which the photoreceptors are arranged. The rods and cones are actually on the inside of the retina (the side toward the lens) but face away from the lens. Thus light must pass through them and bounce off the choroid layer of the retina back toward the lens before it is detected by the rods and cones! It is imperative that the rods, cones and other associated nerve cells in the retina be absolutely optically clear so that distortion of the visible light rays entering the eye and passing through them is minimal.

Many mammals have a special coating on their choroid layer known as the **tapetum lucidum** which gives these mammals their traditional "eye shine" when spotted at night in the headlights of a car. This special layer increases the light gathering ability of the eye and endows these mammals with better night vision. Humans and pigs lack a tapetum lucidum and therefore do not demonstrate this type of eye shine. At one point on the retina you should be able to distinguish the position where the optic nerve exits the back of the eye. This spot is called the **optic disk**. There are no photoreceptors on the surface of the retina at this point, and this confluence of nerves is responsible for the visual phenomenon known as the "blind spot." The opening in front of the lens is known as the **pupil**. A thin sheet of tissue suspended between the cornea and the lens surrounds this opening. This is the **iris** which contains smooth sphincter muscles which contract to decrease the size of the pupil opening and consequently reduce the amount of light which enters the eye. The chamber between the iris and cornea is called the **anterior chamber** and is filled with a liquid called the **aqueous humor**.

ENDOCRINE SYSTEM

Laboratory Objectives

After completing this chapter, you should be able to:

1. Identify the major endocrine organs of the fetal pig.

2. Identify the hormones produced by each endocrine gland and their functions in the body.

3. Identify the microanatomy of the endocrine glands.

The endocrine system is responsible for producing and secreting hormones directly into the bloodstream. **Hormones** are chemical compounds that interact with target cells in the body to produce a myriad of behavioral, neurological and physiological responses. In this way, they influence many of the same behaviors and processes that the nervous system regulates. However, due to the nature of hormones, the effects produced by the endocrine system are generally not short-lived. Nervous responses degrade immediately, but hormones circulating through the bloodstream take anywhere from minutes to hours to breakdown. Thus, hormonal effects tend to be much longer in duration.

You have already identified some of the organs discussed in this chapter. That is because organs that function in the endocrine system may also have other tissue in them that functions in the digestive system (e.g., pancreas) or reproductive system (e.g., ovaries and testes). The centralized control center of the endocrine system is the **hypothalamus-pituitary** complex of the brain. (You will not be able to see this region in your dissection.) Associated with the pituitary gland is the **pineal body**, a small bulb located on the dorsal surface of the forebrain which produces **melatonin**, a hormone which regulates body functions related to the seasonal day length. This hypothalamus-pituitary complex produces many hormones which, in turn, stimulate the activity of many of the other endocrine glands in the body. Likewise, other endocrine organs may produce hormones that stimulate or inhibit regions of the pituitary gland or hypothalamus. By this feedback mechanism, the endocrine system is able to turn itself on and off in response to environmental stimuli.

Neck Region

Examine the ventral aspect of the neck region of your pig. Along either side of the trachea, identify the sections of the **thymus gland** (Figure 9.1). This gland is much larger in the fetal pig (relative to the size of the body) than it will be in the adult. This is because the thymus gland produces **thymosin**, a hormone that stimulates the development of the immune system. Between the lobes of thymus gland, on the ventral side of the trachea, locate the spherical **thyroid gland**. This structure should be quite prominent (it is located a few centimeters caudal to the larynx). The thyroid gland produces two hormones: **thyroxine** which controls the growth rate and metabolic rate of the organism and **calcitonin** which lowers the organism's blood calcium levels.

Abdominal Region

Locate the **pancreas** (identified earlier as a digestive gland) (Figure 9.1). The pancreas also functions in the endocrine system by producing **insulin** and **glucagon** which lower and raise blood glucose levels, respectively, and **somatostatin** which regulates the levels of insulin and glucagon in the blood. Insulin acts primarily on the liver, stimulating it to store more glucose in the form of glycogen, and to a lesser degree on the individual cells of the body, promoting a higher degree of glucose usage. Glucagon works as an antagonist to insulin and reverses the body's actions in these areas. Somatostatin inhibits the release of both insulin and glucagon by the pancreas. There are specific regions of the pancreas known as **islets of Langerhans** which release the hormones into the bloodstream through tiny openings that merge with blood vessels running through the pancreas. On the cranial margin of each kidney (near the midline of the body), a small, lobe-shaped gland is located. These are the **adrenal glands** which regulate blood pressure, carbohydrate metabolism and protein metabolism, and mediate an organism's response to stressful situations. Special cells, known as **interstitial cells**, in the **testes** of the male pig (identified earlier) produce the hormone testosterone. Thus the testes are considered part of the endocrine system. **Testosterone** is responsible for the development and maintenance of the male sexual characteristics and sex drive and the regulation of sperm production. In females, the **ovaries** contain different types of hormone-producing tissues. When the oocyte has matured and ovulation is about to take place, **estrogen** levels rise triggering ovulation and the thickening of the uterine lining. Shortly after ovulation, the remnant tissue from which the oocyte erupted turns into the **corpus luteum** and begins to produce elevated levels of **progesterone**, the hormone that is responsible for increasing the thickness of the endometrial lining. As the levels of these two hormones decrease, the corpus luteum disintegrates and triggers the onset of menstruation.

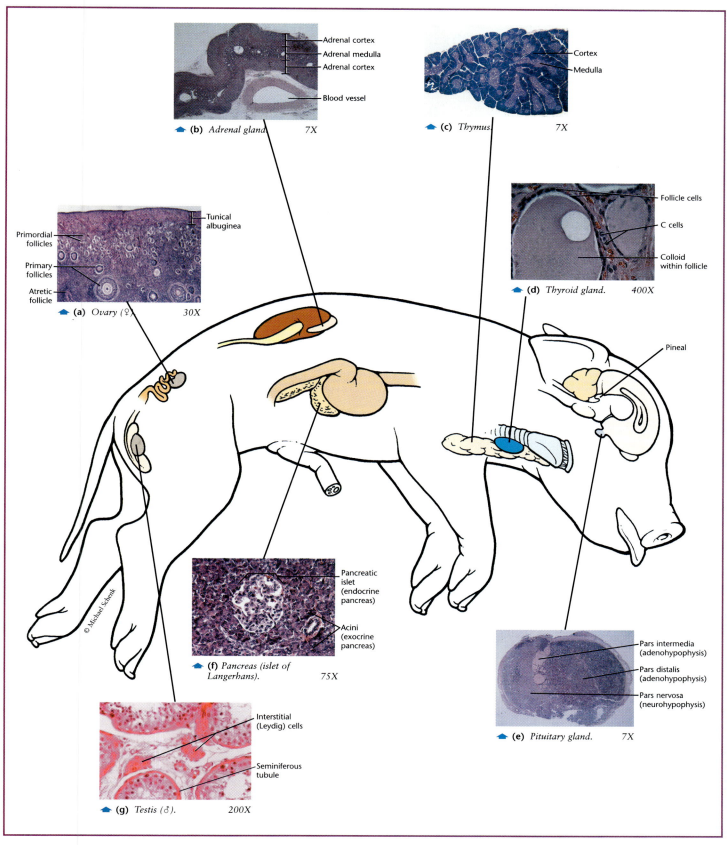

Adrenal cortex
Adrenal medulla
Adrenal cortex
Blood vessel
(b) *Adrenal gland* 7X

Cortex
Medulla
(c) *Thymus* 7X

Tunical albuginea
Primordial follicles
Primary follicles
Atretic follicle
(a) *Ovary (♀)* 30X

Follicle cells
C cells
Colloid within follicle
(d) *Thyroid gland.* 400X

Pineal

Pancreatic islet (endocrine pancreas)
Acini (exocrine pancreas)
(f) *Pancreas (islet of Langerhans).* 75X

Interstitial (Leydig) cells
Seminiferous tubule
(g) *Testis (♂).* 200X

Pars intermedia (adenohypophysis)
Pars distalis (adenohypophysis)
Pars nervosa (neurohypophysis)
(e) *Pituitary gland.* 7X

© Michael Schenk

FIGURE 9.1 *Illustration of endocrine glands in the pig depicting their approximate locations with corresponding histology photographs of each organ.*

TABLE 9.1   *Endocrine glands of the fetal pig, hormones produced by them and their functions in the body.*

Endocrine Gland	Hormone Produced	Hormone Function
Hypothalamus-pituitary complex	Wide array of hormones	Regulate other endocrine glands and control a variety of body functions.
Pineal body	Melatonin	Controls body functions related to photoperiod and seasonal day length and influences sexual maturation.
Thymus	Thymosin	Stimulates immune system.
Thyroid	Thyroxine	Controls metabolism and growth rates.
	Calcitonin	Lowers blood calcium levels.
Pancreas	Insulin	Lowers blood glucose levels.
	Glucagon	Raises blood glucose levels.
	Somatostatin	Inhibits release of insulin and glucagon.
Adrenal	Epinephrine and norepinephrine	Mediate responses to stressful situations.
	Corticosteroids	Control carbohydrate and protein metabolism.
	Aldosterone	Controls blood pressure.
Testes (male)	Testosterone	Maintains male sexual characteristics, sperm production and sex drive.
Ovaries (female)	Estrogen	Induces maturation of oocytes and ovulation; initiates thickening of uterine lining.
	Progesterone	Increases thickening of uterine lining; causes negative feedback which promotes disintegration of corpus luteum.

REFERENCES

Bohensky, F. 1978. *Photo Manual and Dissection Guide of the Fetal Pig.* Avery Publishing Group: Wayne.

Campbell, N. A. 1996. *Biology* (4th ed.). Benjamin/Cummings: Menlo Park.

Chiasson, R. B. and T. O. Odlaug. 1995. *Laboratory Anatomy of the Fetal Pig* (10th ed.). Wm. C. Brown: Dubuque.

Dyce, K. M., W. O. Sack, and C. J. G. Wensing. 1987. *Textbook of Veterinary Anatomy.* W. B. Saunders: Philadelphia.

Evans, H. E. and G. C. Christensen. 1979. *Miller's Anatomy of the Dog* (2nd ed.). W. B. Saunders: Philadelphia.

Fox, S. I. and K. M. Van De Graaff. 1986. *Laboratory Guide to Human Anatomy and Physiology: Concepts and Clinical Applications (Fetal Pig Version).* Wm. C. Brown: Dubuque.

Getty, R. 1975. *Sisson and Grossman's The Anatomy of the Domestic Animals* (5th ed.). W. B. Saunders: Philadelphia.

Hopkins, P. M. and D. G. Smith. 1997. *Introduction to Zoology: A Laboratory Manual* (3rd ed.). Morton Publishing: Englewood.

Rust, T. G. 1986. *A Guide To Anatomy and Physiology Lab* (2nd ed.). Southwest Educational Enterprises: San Antonio.

Van De Graaff, K. M. and J. D. Crawley. 1996. *A Photographic Atlas for the Anatomy and Physiology Laboratory* (3rd ed.). Morton Publishing: Englewood.

Walker, W. F. Jr. 1987. *Functional Anatomy of the Vertebrates: An Evolutionary Perspective.* Saunders College Publishing: Philadelphia.

Walsh, E., K. E. Malone and J. M. Schneider. 1992. *Laboratory Manual for Human Anatomy and Physiology: Fetal Pig Version.* West Publishing Company: St. Paul.

GLOSSARY

–A–

abdomen — region of the body between thorax and pelvis that contains the viscera.

abduct — to move away from the median plane of the body.

adduct — to move toward the median plane of the body.

adrenal gland — endocrine gland located on medial side of kidney that produces hormones which mediate responses to stressful situations and control blood pressure and carbohydrate and protein metabolism.

aldosterone — hormone produced by adrenal gland that controls blood pressure by acting on the reabsorption of sodium ions by the kidney and regulating water flow into the kidney.

allantoic duct — tube passing through the umbilical cord of the fetus connecting it with the allantois in the uterus of the mother.

allantois — extra-embryonic sac that acts as a repository for metabolic wastes produced by the fetus during development.

alveoli — (singular = alveolus); multilobed air sacs that form the terminal ducts of the bronchioles of the lungs and serve as the surfaces for the exchange of carbon dioxide and oxygen.

amylase — enzyme component of saliva that breaks down starches.

anterior chamber — fluid-filled region of the eye located between the cornea and the iris.

anus — opening of the rectum through which undigested food particles (feces) are egested from the body.

aorta — large artery arising from the left ventricle that distributes blood to the regions of the body.

appendicular skeleton — portion of the skeletal system consisting of the pectoral and pelvic girdles and the forelimbs and hindlimbs.

aqueous humor — viscous liquid component of the anterior chamber of the vertebrate eye.

artery — blood vessel that carries blood away from the heart.

articulation — juncture between two or more bones (usually a movable joint).

atlas — (1) the first cervical vertebra; modified for attachment with the skull. (2) a collection of illustrations, tables and text providing information on a topic.

atrium — (plural = atria); chamber of the heart that receives blood.

axial skeleton — portion of the skeleton consisting of the skull, vertebral column and rib cage.

axis — (1) the second cervical vertebra. (2) a straight line that bisects the body into two equal halves; usually along the longer portion of the body.

–B–

bicuspid valve — (also called mitral valve); valve of the mammalian heart that directs blood flow from the left atrium to the left ventricle; so named because it has two cusps.

bile — digestive fluid secreted by liver and stored in gallbladder; functions in the emulsification of fats in the duodenum.

bone — rigid connective tissue used to support the body; characterized by densely packed, hard fibrous matrix composed of calcium salts surrounding osteocytes (bone producing cells).

Bowman's capsule — cup-shaped layer of epithelial tissue surrounding the glomerulus of the vertebrate nephron which receives the filtrate of the blood.

brachiocephalic trunk — major branch of the aorta that supplies blood to the head and upper trunk region of the body.

brain — part of the central nervous system responsible for processing and integrating nerve impulses gathered from all sensory organs and receptors throughout the body.

bronchi — (singular = bronchus); major divisions of the trachea that supply oxygen (and remove carbon dioxide) from the lobes of the lungs.

bronchiole — finer subdivision of the bronchi that forms a branching arrangement and carries gases to and from the regions within the lobe of a lung.

bulbourethral glands — accessory glands of the male reproductive system located at the base of the penis and urethra that contribute fluid to the semen which aids in neutralizing the acidity of the vagina.

–C–

calcitonin — thyroid hormone responsible for lowering blood calcium levels.

cardiac muscle — type of muscle tissue that forms the walls of the heart; characterized by striated muscle fibers joined together with gap junctions called intercalated disks which relay each heartbeat.

cardiovascular — of or pertaining to the heart and vascular system.

cartilage — flexible connective tissue that is characterized by fibrous tissue surrounding individual chondrocytes (cartilage-producing cells).

caudal — situated more toward the posterior (tail) region of the body.

caudal vena cava — the major vein returning deoxygenated blood from the lower extremities of the body to the right atrium of the mammalian heart.

cecum — blind projection located at the junction of the ileum and colon that serves as a sac where fermentation of cellulose occurs. The cecum plays a prominent role in the digestive process of most herbivores, but is reduced in omnivores and carnivores.

central nervous system — portion of the nervous system consisting of the brain and spinal cord.

cerebellum — region of the vertebrate hindbrain that integrates the movements of skeletal muscles and controls coordination and balance.

cerebrum — the portion of the brain devoted to the integration of sensory impulses, learning, memory and voluntary movements; divided into two hemispheres and located in the upper portion of the cranial cavity.

cervix — constricted portion of the female reproductive tract between the opening to the uterus and the vagina.

chordae tendineae — tendinous fibers connecting the valves of the mammalian heart to the papillary muscles associated with the ventricles of the heart.

choroid layer — vascular coating of the eye located between the sclera and the retina.

chyme — fluid produced by the action of digestive enzymes from the stomach mixing with and dissolving ingested food particles.

ciliary body — small muscles associated with the lens in the vertebrate eye; responsible for changing the shape of the lens to focus on objects at different distances.

collecting duct — tubule of the mammalian kidney that receives filtrate from the convoluted tubules and loop of Henle and sends it to the ureter for transport out of the kidney; allows water to be reabsorbed by bloodstream producing a highly concentrated urine.

colon — portion of the large intestine extending from the cecum to the rectum that functions primarily in reabsorbing water that has been added during the digestive process.

common bile duct — tubule through which bile is transported from the liver to the gallbladder and from the gallbladder to the duodenum.

cone — photoreceptor located in the mammalian eye that detects color.

convoluted tubules — region of the mammalian nephron that permits reabsorption of water and salts by the bloodstream.

cornea — transparent outer layer of the eye.

coronary artery — one of several small arteries located on the ventral surface of the heart that supply freshly-oxygenated blood to the tissue of the heart.

corpus luteum — region of the mammalian ovary that forms after the mature oocyte has erupted from the ovary; produces progesterone.

cortex — outer region of an organ; "renal cortex" refers to outermost layer of kidney.

corticosteroids — hormones produced by adrenal glands which control protein metabolism and carbohydrate metabolism.

cranial — situated toward the head region.

cranial vena cava — the major vein returning deoxygenated blood from the upper extremities of the body to the right atrium of the mammalian heart.

cremaster muscles — small muscles attached to the testes that retract the testes toward the abdominal cavity; function to keep testes at a constant temperature by controlling their proximity to the body wall.

cystic duct — tubule connecting pancreas to the duodenum; digestive enzymes produced by pancreas are secreted into the duodenum through this canal.

–D–

diaphragm — muscular sheet separating the thoracic and abdominal cavities; used to ventilate the lungs of mammals.

digestion — process by which ingested food particles are broken down into smaller units that can be utilized by individual cells in the body.

digitigrade — type of locomotion characterized by walking on the tips of the toes (digits); body weight is supported primarily by the phalanges.

dissection — the process or act of uncovering and exposing tissues and organs of an animal by teasing apart or cutting structures.

distal — situated toward the outer extremity of the body, away from the median plane (e.g., your hand is distal to your shoulder).

dorsal — situated toward the back of the body, closer to the vertebral column.

dorsal aorta — descending portion of the aorta that runs caudally along the ventral surface of the vertebral column and carries oxygenated blood from the left ventricle to the caudal regions of the body.

ductus arteriosus — short connection joining the pulmonary trunk with the aorta and allowing a portion of the blood from the pulmonary trunk to enter the aorta instead of flowing to the lungs; found only in the fetus.

ductus deferens — tube connected to the epididymis that transports sperm from the testis and epididymis to the urethra during ejaculation.

duodenum — first portion of the small intestine; functions primarily in final stages of chemical digestion and begins process of nutrient absorption.

–E–

ears — external sensory receptors that pick up airborne vibrations and send them to the brain where they are interpreted as sounds.

egestion — the process of expelling undigested food particles through the anus.

endocrine — pertaining to the endocrine system — system responsible for the production of hormones that communicate chemically with target organs through the bloodstream.

endoskeleton — a hard skeleton used for support that is embedded within the soft tissues of the body.

endothermy — condition in which an animal uses its own metabolic processes to maintain a constant internal body temperature.

epididymis — highly coiled tubule system that cups around the testis and serves as a storage unit and transportation canal for mature sperm.

epiglottis — cartilaginous flap that covers the glottis to prevent food from entering the larynx and trachea when swallowing.

epinephrine — (also called adrenaline); hormone produced by adrenal gland that causes the body to respond in stressful situations.

esophagus — muscular passageway connecting the mouth and oral cavity to the stomach.

estrogen — primary ovarian hormone produced by the follicle that stimulates the development and maintenance of female reproductive system and secondary sexual characteristics.

excretion — process of eliminating metabolic waste products produced through cellular metabolism from the body.

exocrine — referring to tissues not associated with the endocrine system; usually non-hormone producing glands or organs that are in close proximity to endocrine tissues.

exoskeleton — hard outer skeleton covering the body of an animal, such as the cuticle of arthropods or the shell of molluscs.

extensor — any muscle that extends a limb or joint through contraction.

eyes — external sensory receptors that receive light rays and convert them into neural impulses which are sent to the brain and interpreted as vision.

–F–

fascia — thin sheet or band of fibrous connective tissue that binds tissues or organs together and holds them in place.

feces — excrement produced by the digestive process that is eliminated through the anus.

flexor — any muscle that draws a limb or joint closer to the axis of the body.

follicle — a structure within the ovary that contains the developing oocyte.

foramen ovale — opening between the right and left atria in the fetal heart that allows a mixing of blood between these two chambers.

frontal — situated toward the ventral half of the body; denoting a longitudinal plane.

–G–

gallbladder — organ located on the underside of the liver which stores bile and releases it into the duodenum.

gamete — reproductive cell produced in the gonads through meiosis; a haploid egg or sperm cell.

genital papilla — small, fleshy projection around the urogenital opening of the female fetus which is homologous to the penis in the male.

gestation — the period of embryonic development from the time of fertilization to birth in viviparous (live-bearing) species.

glomerulus — capillary bed of the nephron that filters out fluids and small chemical particles from the blood into the surrounding Bowman's capsule.

glottis — opening in the oral cavity that leads from the nasopharynx to the larynx and trachea.

glucagon — pancreatic hormone that raises blood glucose levels.

glycogen — converted form of glucose that is stored in the liver and muscles of animals.

–H–

hard palate — bony plate separating the rostral portion of the oral cavity from the nasopharynx in mammals.

head — region of the body in mammals consisting of the skull, brain and major sense organs.

hepatic portal system — system of blood vessels that carries blood from the capillary beds of the stomach, small intestines and spleen to another capillary bed in the liver, where blood is detoxified and nutrients are stored and released at a controlled rate.

hepatic portal vein — large vessel that carries nutrient-rich and toxin-laden blood from the small intestines and pancreas to the liver for detoxification and regulation of nutrient release before the blood passes to the rest of the body.

homologous structures — structures in different species that are similar due to common ancestry shared by the species.

hormone — chemical compound produced by endocrine tissue and distributed through the body via the circulatory system that communicates with target organs and tissues to produce a wide array of behavioral and physiological responses, depending on the specific hormone released.

hydrochloric acid — one of the major constituent chemicals released by the stomach as a digestive compound.

hypothalamus — region of the brain responsible for coordinating the efforts and integration of the endocrine and nervous systems; produces a wide range of hormones.

–I–

ileum — distal portion of the small intestine extending from the jejunum to the cecum; primarily responsible for absorption of nutrients.

ilium — broad, flat, uppermost region of the pelvis; ilium is fused with ischium and pubis to form the pelvis.

ingestion — the process of taking in food through the oral cavity.

insertion — the distal point of attachment of a muscle, usually to the bone moved by that muscle.

insulin — hormone secreted by the endocrine cells of the pancreas (islets of Langerhans) that is responsible for lowering blood glucose levels by stimulating the liver to store more glucose as glycogen.

interstitial cells — hormone-producing cells situated between the seminiferous tubules of the testes that produce testosterone.

iris — region of the eye that regulates the amount of light that enters the eye and reaches the retina by contraction of the sphincter muscles of the iris.

–J–

jejunum — middle portion of the small intestine extending from the duodenum to the ileum; primarily responsible for nutrient absorption.

–K–

kidney — excretory unit located in the lumbar region of mammals; this structure filters the blood creating a highly-concentrated metabolic by-product (urine) which is sent to the urinary bladder; also responsible for maintaining a homeostatic balance of salts, fluids and ions within the body (osmoregulation).

–L–

larynx — enlarged, oval-shaped region cranial to the trachea that contains the vocal cords.

lateral — situated farther away from the midline (median plane) of the body.

lens — biconvex structure in the vertebrate eye located behind the iris; functions to focus images on the retina.

liver — large, multilobed organ of the abdominal cavity located just caudal to the diaphragm; secretes bile, filters toxins and nutrients from the blood and stores sugars.

longitudinal fissure — crevice running down the median plane of the cerebrum separating the brain into left and right hemispheres.

loop of Henle — long projection of the tubules of the nephron that descends into the medulla of the kidney; creates a concentration gradient that allows salts and water to be reabsorbed by the body, while nitrogenous wastes are retained in the nephron and concentrated.

lumbar — pertaining to the lower back region of the body.

–M–

mammal — class of vertebrates characterized by animals that bear live young (typically), provide milk for young from mammary glands, possess fur or hair and have a single lower jaw bone (the dentary bone).

mammary glands — modified tissues on the ventral surface of mammals that secrete milk to nourish their young.

mammary papillae — small protuberances on the ventral surface of the fetal pig; in adult females the papillae will develop into teats through which the mammary glands secrete milk.

medial — situated toward the midline of the body.

median plane — longitudinal section running down the exact midline of a bilaterally symmetrical animal.

medulla — middle region of the kidney; contains loops of Henle and some collecting ducts.

medulla oblongata — most caudal region of the vertebrate brain; controls autonomic functions such as breathing, heart rate, digestion and swallowing.

meiosis — process of cell division whereby a diploid cell undergoes reduction division and results in four haploid daughter cells, typically referred to as gametes.

melatonin — pituitary hormone which influences sexual maturation and controls body's responses to seasonal changes in day length.

mitral valve — (also called bicuspid valve); valve of the mammalian heart that directs blood flow from the left atrium to the left ventricle.

mesentery — connective membrane that suspends body organs in the abdominal cavity and holds them together.

muscle — a type of tissue specialized for creating movement through contractions of the individual fibers that make up the tissue; designed to either move an animal through its environment or move substances through the animal.

–N–

nares — the external openings of the nasal passageway; utilized in respiration.

nasopharynx — region of the nasal passageway above the soft palate.

nephron — functional unit of the kidney; specialized subunit that filters blood and concentrates urine.

norepinephrine — adrenal hormone that mediates an animal's responses to stressful situations.

–O–

occipital region — (also called occipital lobe); posterior portion of the cerebrum where the optic lobes are located.

oocyte — (also called ovum); an immature egg produced in the ovary.

optic disk — region of the vertebrate eye where the neurons of the optic nerve pass through the choroid layer and retina; commonly referred to as the "blind spot" since there are no visual receptors in this area.

optic nerve — large confluence of nerve fibers from all the photoreceptors of the eye.

origin — the less movable anchor point of a muscle attachment.

ovary — reproductive organ in females that produces eggs and hormones.

oviduct — tube through which the egg, upon leaving the ovary, is carried on its way to the uterine horns.

ovulation — process by which mature eggs are released from the ovaries; characterized by a surge in hormone levels and a corresponding thickening of the uterine lining.

–P–

pancreas — granular organ located along the left margin of the duodenum and the caudal margin of the stomach; produces digestive enzymes and a variety of hormones.

pancreatic duct — canal through which digestive enzymes produced by the pancreas are transported to the duodenum.

papilla — small nipple-shaped projection or elevation.

parietal region — lobe of the cerebrum located on either side of the head near the base of the skull.

parotid duct — small canal leading from parotid gland to oral cavity through which the parotid gland releases its salivary enzymes into the mouth.

parotid gland — rather large salivary gland located near the ear of the pig.

penis — external reproductive organ of the male; deposits semen in the reproductive tract of the female and carries excretory wastes in the form of urine out of the body through the urethra.

pepsinogen — gastrointestinal compound secreted by the gastric cells of the stomach that is instrumental in the chemical digestion of food particles.

pericardial membrane — thin tissue surrounding and protecting the heart.

peripheral nervous system — compilation of sensory and motor neurons and nerve fibers associated with the forelimbs and hindlimbs of the body.

peristalsis — rhythmic contractions of the alimentary canal which propel food along its length.

pineal body — small protuberance of the cerebrum which secretes hormones into the bloodstream.

pituitary gland — endocrine gland located at the base of the hypothalamus that directs the functions of many other endocrine glands throughout the body.

plantigrade — type of locomotion characterized by walking on the soles of the feet; body weight is supported primarily by the tarsals.

progesterone — hormone produced by the corpus luteum of the ovary that is responsible for preparing the uterus for reception and development of the fertilized eggs.

proximal — situated toward the trunk of the body, closer to the median plane (e.g., your elbow is proximal to your hand).

pulmonary arteries — short blood vessels which, in the adult, carry deoxygenated blood from the right ventricle of the heart to the lungs.

pulmonary veins — blood vessels which, in the adult, carry oxygenated blood from the lungs to the left atrium of the heart.

pupil — opening in the iris of the eye; its size is controlled by contractions of the sphincter muscles of the iris to regulate the amount of light that enters the eye.

–Q–

quadrupedal — describes an animal that walks on all four legs.

–R–

rectum — most distal end of the intestinal tract; primary responsibility is to reabsorb water and produce dry, concentrated feces.

renal pelvis — innermost region of the kidney; contains the collecting ducts and the origin of the ureter.

retina — specialized layer of the vertebrate eye that contains the photoreceptive cells (the rods and cones).

rod — type of photoreceptor that "sees" images as only black and white; these cells are very good at detecting motion and have extremely high visual acuity.

rostral — situated toward the tip of the nose.

rugae — ridges and folds of the inner wall of the stomach which increase the surface area of the stomach lining and provide texture for the manipulation of food as it is broken down.

–S–

sacral — pertaining to the sacrum.

sacrum — wedge-shaped portion of the pelvis that is formed by the fusion of 4 vertebrae (in the pig) and serves to support the pelvic girdle and hindlimbs.

sagittal — refers to a plane running the length of the body parallel to the median plane.

saliva — liquid secretion of the salivary glands that lubricates food to facilitate swallowing and contains enzymes which initiate the digestive process.

salivary glands — special glands located along the oral cavity and neck that produce a variety of fluids and enzymes that facilitate digestion.

sclera — tough, outer covering of the eye; gives the outer eyeball its white coloration; protects the delicate inner structures and serves as a tissue for muscle attachments.

scrotal sacs — pouches extending from the caudal region of the pig which will contain the testes after they have descended from the abdominal cavity. Their presence allows the temperature of the testes to be maintained at a slighter lower temperature than that of the abdominal cavity.

secondary palate — region that comprises the "roof of the mouth," separating the nasal passageway from the oral cavity; in mammals it is comprised of the hard and soft palates.

semen — mixture containing sperm cells and accessory fluids secreted by the reproductive glands of the male; serves to provide a nutrient base for sperm as well as keep them moist and neutralize the acidity of the vagina to increase sperm survival.

semilunar valve — flaps of tissue at the junction of each ventricle of the heart to prevent backflow of blood from either the pulmonary arteries or aorta into their respective ventricles.

seminal vesicles — accessory glands of the male reproductive system located near the juncture of the urethra and base of the penis; they contribute fluid to the semen that contains nutrients for the sperm and stimulates uterine contractions to assist in directing the sperm toward the egg.

seminiferous tubules — tubule system located inside the testes where sperm are produced through meiosis. Primary spermatocytes are located along the outer margins of the seminiferous tubules and move inward as they mature.

sensory neuron — specialized nerve cell that is capable of receiving external stimuli and sending a nerve impulse through the nervous system to the spinal cord and brain.

skeletal muscle — type of muscle tissue characterized by striated fibers and multinucleated cells; typically under voluntary control.

skull — hard, bony, protective covering of the brain.

smooth muscle — type of muscle tissue characterized by fibers with no striations and a single nucleus in each muscle cell; typically involuntary.

soft palate — cartilaginous region of the roof of the mouth that separates the oral cavity from the nasal passageway; located toward the back of the mouth.

somatostatin — pancreatic hormone that regulates the levels of insulin and glucagon in the blood.

spinal cord — thin extension of the central nervous system that runs along the length of the body, protected by the bony vertebrae.

spleen — dark, oblong organ in the abdominal cavity that is a component of the circulatory system; recycles old red blood cells, stores them and releases them into circulatory system.

stomach — large U-shaped digestive reservoir for food. In addition to storing large quantities of food, chemicals are secreted by the walls of the stomach which break the food down into microscopic particles that may be absorbed by the cells of the intestines.

sublingual gland — salivary gland located underneath the skin and alongside the tongue of the pig.

submaxillary gland — oval-shaped salivary gland located underneath the large parotid gland.

superficial — lying near the surface.

–T–

tactile — relating or pertaining to the sense of touch.

tapetum lucidum — reflective coating of the choroid layer of the eye of some mammals which increases their ability to see at night and is responsible for the phenomenon of "eye shine."

tendon — fibrous cord of connective tissue which typically serves as an attachment between a muscle and bone.

testis — reproductive organ of the male which produces sperm and hormones.

testosterone — the principal male sex hormone; responsible for the development and maintenance of male secondary sexual characteristics and sex drive.

thoracic — pertaining to the chest region.

thorax — region of the body from the base of the neck to the point where the diaphragm extends across the body cavity.

thymosin — hormone produced by the thymus gland which stimulates the action of the immune system.

thymus — endocrine gland located along the lateral margins of the trachea near the larynx and lying on the cranial margin of the pericardial membrane surrounding the heart; produces thymosin.

thyroid — oval-shaped endocrine gland located on the ventral surface of the trachea just caudal to the larynx; produces thyroxine and calcitonin.

thyroxine — thyroid hormone responsible for controlling metabolic and growth rates.

tongue — muscular structure located in the oral cavity and used for manipulation of food.

trachea — cartilaginous tube extending from larynx to the lungs through which air is transported during respiration.

transverse — referring to a plane separating the body into cranial and caudal portions (perpendicular to the median plane).

tricuspid valve — flaps of tissue at the juncture of the right atrium and right ventricle which prevent backflow of blood into the right atrium.

trunk — region of the body extending from the point where the diaphragm bisects the body to the base of the tail.

–U–

umbilical cord — attachment between the maternal placenta and the fetus through which gases, nutrients and nitrogenous wastes are transported during embryonic development.

ureter — tube that transports urine from the kidney to the urinary bladder for storage.

urethra — tube that leads from the urinary bladder through the penis to the outside of the body; transports urine and (in males) semen.

urinary bladder — membranous sac that serves as a receptacle for excreted urine from the kidneys.

urine — fluid excreted by the kidneys, stored in the urinary bladder and eliminated from the body through the urethra; composed primarily of nitrogenous wastes and excess salts and sugars.

urogenital opening — opening of the urethra (in males) or the urogenital sinus (in females) through which urine passes as it is eliminated from the body.

urogenital sinus — common chamber for reproductive and excretory functions in the female; located just caudal to the junction of the vagina and urethra.

uterus — region where embryonic development of the fetus occurs; in pigs the uterus is divided into the body of the uterus and two uterine horns. The uterine horns are where development occurs in the pig.

–V–

vagina — female reproductive canal situated between the urogenital sinus and the cervix.

vein — blood vessel that carries blood toward the heart.

ventral — situated toward the belly region of an animal.

ventricle — large muscular chamber of the heart that pumps blood out of the heart into an artery.

vertebrate — animal that possesses bony vertebrae that surround the spinal cord.

vibrissae — hairs that project outward from the head of an animal and respond to tactile stimuli (often called whiskers).

vitreous chamber — posterior fluid-filled chamber of the eye that contains the lens.

vitreous humor — clear, jelly-like liquid that fills the vitreous chamber; provides support and cushioning for the lens and internal structures of the eye.

INDEX

A

Abdomen, 5, 15, 22
Abdominal arteries, 63
Abdominal cavity, 34–41, 53–64
Abdominal region, 58–60, 92
Abdominal veins, 63
Abdominal vessels, 58
Abducens nerve (VI), 85
Abducts, 13
Accessory duct, 38
Accessory lobe, 46, 66, 67
Accessory pancreatic duct, 39
Acini, 40, 93
Adductor, 25, 26, 27, 28
Adduct, 13
Adrenal, 58, 81
Adrenal cortex, 93
Adrenal glands, 59, 63, 72, 74, 77, 79, 92, 93, 94
Adrenal medula, 93
Afferent arteriole, 80, 81, 82
Allantoic duct, 60
Allantoic stalk, 47
Allantois, 15
Alveolar duct, 68
Alveolar sacs, 68
Alveoli, 67, 68
Amylase, 33
Anatomical planes, 2
Ankle, 5, 6, 7
Anterior chamber 88, 89
Anus, 6, 15, 41, 74, 77
Aorta, 46, 47, 48, 54, 57, 58, 59, 61, 63, 75, 77, 79
Aortic arch, 44, 46, 48, 52, 55
Aortic semilunar valve, 53
Apical lobe
  cranial segment, 66
  medial segment, 66
Aponeurosis of biceps femoris, 28
Aponeurosis of gracilis, 25, 26
Appendicular skeleton, 10, 11
Aqueous humor, 89
Arcuate vessels, 81
Arterial arcades of the mesentery, 59
Arterial supply, 58
Artery(ies), 44, 49, 53, 57
  aorta, 46, 47, 48, 54, 57, 58, 59, 61, 63, 75, 77, 79

axillary, 21, 49, 51, 52
brachial, 51
brachiocephalic trunk, 46, 48, 49, 51, 52, 55
carotid, common, 48, 51, 52
carotid, external, 51, 52
carotid, internal, 51, 52
caudal mesenteric, 58, 60, 63
celiac, 58
coronary, 44, 45, 46, 54
cranial mesenteric, 38, 57, 58, 59, 63
dorsal aorta, 51
ductus arteriosus, 46, 47, 48, 54, 55
femoral, 58, 60, 61, 63
genital, 60
iliac, external, 58, 60, 61, 63
iliac, internal, 58, 60, 61, 63
pulmonary, 44, 45, 46, 47, 48, 52, 55, 66
renal, 58, 59, 60, 61, 63, 72, 78, 79, 81, 82
subclavian, 46, 48, 49
thoracic, internal, 51, 52
umbilical, 15, 35, 36, 47, 58, 60, 61, 62, 72, 74, 75, 76, 77
Arthropods, 9
Ascending colon, 40, 41
Atlas, 10, 11
Atretic follicle, 93
Auditory nerve (VIII), 85
Auricle, 46, 48
Axial skeleton, 9, 10, 11
Axillary artery, 21, 49, 51, 52
Axillary vein, 21, 49, 50
Axis, 10, 11
Azygos vein, 54, 55

B

Ball and socket joint, 12
Basement membrane, 68
Biceps, 16, 21, 22
Biceps femoris, 18, 19, 22, 24, 28
Bicuspid valve, 53, 54, 55
Bile, 37
Bile duct, 37
Blood pressure, 92
Blood vessel, 93
Body (of the uterus), 75, 77
Body planes, 2
Body regions, 2, 3

Bone(s)
  atlas, 10, 11
  axis, 10, 11
  calcaneus, 10, 11
  carpal(s), 10, 11
  femur, 10, 11
  fibula, 10, 11
  frontal, 10, 11
  humerus, 10, 11
  ilium, 10, 11, 40, 41
  ischium, 10, 11
  mandible, 10, 11, 16, 18, 69
  maxilla, 10, 11
  metacarpal(s), 10, 11
  metatarsal(s), 10
  nasal, 10, 11
  occipital, 10, 11
  parietal, 10, 11
  patella, 10, 11
  phalange(s), 10, 11
  premaxilla, 10, 11
  radius, 10, 11
  ribs, 10, 11
  sacrum, 10, 11
  scapula, 10, 11
  tarsal(s), 10, 11
  temporal, 10, 11
  tibia, 10, 11, 25
  ulna, 10, 11, 19
  vertebra, 9, 10
  zygomatic arch, 10, 11
Bowman's capsule, 80, 81, 82
Brachial plexus, 21, 35, 51
Brachialis, 16, 19, 20
Brachiocephalic artery, 54
Brachiocephalic trunk, 46, 48, 49, 51, 52, 55
Brachiocephalic vein, 49
Brachiocephalicus, 16, 17, 18
Brain, 83–86
Broad ligament, 77
Bronchi, 66
Bronchiole, 46, 66, 67, 68
Bulbourethral glands, 72, 73, 74, 78

C

C cells, 93
Calcitonin, 92
Calcaneus bone, 10, 11
Calyx, 81
Canines, 11, 34
Capillary, 68

Capillary bed, 53
Cardiac muscle, 13, 14
Cardiovascular system, 43–64
Carotid trunk, 51
Carpals, 10, 11
Caudal lobe, 66
Caudal mesenteric artery, 58, 60, 63
Caudal region, 2, 3
Caudal vena cava, 37, 45, 46, 47, 48, 50, 53, 54, 55, 56, 58, 59, 60, 63, 66, 74, 75, 77
Caudal vertebrae, 10, 11
Cecum, 36, 40, 41, 57, 64
Celiac artery, 58
Celiac axis, 57, 63
Central artery, 64
Central artery of retina, 88
Central (haversian) canals, 10
Central nervous system (CNS), 83
Cephalic vein, 49, 50
Cerebellum, 84, 85, 86
Cerebrum, 84, 85, 86
Cervical vertebra, 10, 11
Cervix, 76
Chondrocytes, 10
Chordae tendineae, 53, 54
Choroid layer, 87, 88
Chyme, 38
Ciliary bodies, 87, 88
Circulatory system, 43
Cleidomastoid, 16, 18, 19
Cleidooccipitalis, 16, 18, 19
Collecting duct, 80, 81, 82
Colloid (within follicle), 93
Colon, 35, 36, 41, 59, 60, 64, 75, 76
Common bile duct, 38, 39, 40, 56, 57
Common carotid artery, 48, 51, 52
Common iliac artery, 61
Common iliac vein, 61
Condylar joint, 12
Coracobrachialis, 22
Cornea, 87, 88
Coronary artery, 44, 45, 46, 54
Coronary sinus, 45, 46, 54, 55
Coronary vein, 44
Coronary vessels, 46, 54
Corpus callosum, 86
Corpus luteum, 92

Cortex, 80, 81, 93
Cortical labyrinth, 81
Costocervical artery, 46, 52
Costocervical trunk, 49
Costocervical vein, 46, 50
Cranial lobe, 66, 67
Cranial mesenteric artery, 38, 57, 58, 59, 63
Cranial nerves, 83, 86
Cranial region, 2, 3
Cranial segment, 66
Cranial vena cava, 38, 44, 45, 46, 47, 48, 50, 52, 54, 55, 63
Cremaster pouch, 73
Cystic artery, 37, 57
Cystic duct, 37, 38, 56, 57

D

Deep circumflex artery, 61
Deep circumflex vein, 61
Deep circumflex iliac artery, 58, 63
Deep circumflex iliac vein, 58
Deep femoral artery, 58, 60, 61, 63
Deep femoral vein, 58, 60, 61
Deltoid, 18, 19, 20
Diaphragm, 5, 35, 36, 44, 46, 50, 52, 58, 66
Digastric, 16, 17, 18
Digestive organs, 5
Digestive system, 31–41
Digitigrade, 11
Digits, 7
Dissection techniques, 2
Distal convoluted tubules, 80, 82
Distal region, 2, 3
Distal tubule, 81
Dorsal aorta, 51
Dorsal region, 2, 3
Duct(s)
    accessory, 38
    allantoic, 60
    bile, 37
    cystic, 37, 38, 56, 57
    pancreatic, 38, 39
    parotid, 31, 33
Ductus arteriosus, 46, 47, 48, 54, 55
Ductus deferens, 72, 73, 74, 78, 79
Duodenal ampullae, 39
Duodenum, 37, 39, 40, 41, 57

E

Ears, 5, 6, 15
Efferent arteriole, 80, 81, 82
Elbow, 6, 7, 20
Endoskeleton, 9
Endothermy, 69
Endocrine system, 91–94
Epididymis, 72, 74, 78
Epiglottis, 33, 34, 67, 69
Esophagus, 33, 34, 41, 48, 58, 66, 69
Estrogen, 92
Excretory organs, 5
Excretory system, 78–82

Exoskeletons, 9
Extensor carpi oblique, 19
Extensor carpi radialis, 16, 19, 20, 21
Extensor digiti V
Extensor digitorum communis, 19, 20
Extensor digitorum lateralis, 19, 20
Extensor digitorum longus, 23, 24, 28
Extensor digitorum quarti and quinti, 23, 24, 28, 29
External anatomy, 5–8
External carotid artery, 51, 52
External features, 5–8
External iliac artery, 58, 60, 61, 63
External iliac vein, 58, 60, 61, 63
External jugular vein, 33, 49, 50
External maxillary vein, 33
External oblique, 16, 18, 19, 22, 23, 26
External thoracic artery, 51, 52
External thoracic vein, 50
Eye, 5, 15, 84, 87
Eyelid, 6, 88

F

Facial nerve, 32, 33, 85
Facial vein, 52
Fascial sheath, 74
Female reproductive system, 75–78
Femoral artery, 58, 60, 61, 63
Femoral nerve, 61
Femoral vein, 58, 60, 61
Femur, 10, 11
Fetal circulatory pathway, 47
Fetal pig,
    skeletal development, 10
Fetal skeleton, 9–12
Fetal system, 5
Fibula, 10, 11
Flexor carpi radialis, 16, 20, 21
Flexor carpi ulnaris, 16, 20, 21
Flexor digitorum longus, 23, 25, 26, 28
Flexor digitorum profundus, 16, 19, 20, 21
Flexor digitorum superficialis, 16, 20, 21
Flexor hallicus, 23, 25, 26, 28
Follicle cells, 93
Foramen ovale, 48
Forelimbs, 10
Fourth ventricle, 86
Frontal bone, 10, 11
Frontal lobe, 86
Frontal plane, 2, 3
Frontal region, 84
Fungiform, 34

G

Gallbladder, 36, 37, 40, 41, 56, 57
Gametes, 71

Gastrocnemius, 23, 24, 25, 26, 28, 29
Gastrohepatic artery, 57
Gastrosplenic vein, 56, 57
Genital arteries, 60
Genital papilla, 6, 7, 15, 76, 77
Genital veins, 60
Genital vessels, 58
Germinal center, 64
Gland(s), 40
    adrenal, 58, 81
    bulbourethral, 72, 73, 74, 78
    endocrine, 40
    mammary, 16, 17, 18, 32
    parotid, 31, 33
    pituitary, 85, 86, 91, 93
    salivary, 32, 33, 41
    sublingual, 31, 33
    submaxillary, 31, 33
    thymus, 17, 32, 36, 92, 93, 94
    thyroid, 35, 36, 43, 49, 52, 92, 93, 94
Gliding joint, 12
Glomerulus, 80, 81, 82
Glossopharyngeal nerve (IX), 85
Glottis, 33, 34
Glucagon, 92
Gluteus medius, 18, 19, 23, 24, 28
Gluteus profundus, 27, 28, 29
Gluteus superficialis, 18, 23, 24, 28
Gracilis, 23, 24, 25, 26
Greater trochanter, 28
Growth plates, 10
Gubernaculum, 74
Gyrus, 85

H

Hard palate, 33, 34, 67
Head, 5, 15, 31–33
Heart, 35, 36, 43, 46, 55
Hepatic artery, 57
Hepatic duct, 56
Hepatic portal system, 53, 56, 57
Hepatic portal vein, 37, 38, 39, 53, 56, 57, 63
Hepatic vein, 38, 56
Hindlimbs, 10, 23–29
Hinge joint, 12
Homologous structures, 9
Hormone, 91
Horns (of the uterus), 75, 76, 77
Human eye, 87
Humerus, 10, 11
Hyaline cartilage, 10, 68
Hydrochloric acid, 35
Hypoglossal nerve (XII), 85
Hypothalamus, 91
Hypothalamus-pituitary complex, 91, 94

I

Iliacus, 25, 26, 27, 28
Ilium, 10, 11, 40, 41
Incisors, 11, 34

Insertion, muscle, 13
Insulin, 92
Intercalated discs, 14
Interlobar vessels, 81
Internal carotid artery, 51, 52
Internal iliac artery, 58, 60
Internal iliac vein, 58, 60
Internal jugular vein, 32, 49, 50, 51, 52
Internal maxillary vein, 33
Internal oblique, 18, 22, 23
Internal thoracic artery, 46, 51, 52
Internal thoracic vein, 46, 49, 50
Interstitial cells, 74, 92, 93
Iris, 88
Ischium, 10, 11
Islets of Langerhans, 40, 92, 93

J

Joint(s), 12
    condylar, 12
    gliding, 12
    hinge, 12
    pivot, 12
    spheroidal, 12
    suture (immovable), 12
Jugular vein, 32
Jejunum, 38, 40, 41, 56, 59

K

Kidneys, 7, 58, 59, 61, 63, 72, 74, 77, 78, 79, 81
Knee, 6, 7

L

Lacunae, 10
Lamina propria, 40
Large intestine, 40
Larynx, 16, 17, 35, 65, 66, 67, 69, 92
Lateral plane, 2, 3
Lateral region, 2
Latissimus dorsi, 16, 18, 20
Left atrium, 45, 54, 55
Left auricle, 44, 46, 48, 54, 55
Left axillary artery, 51
Left axillary vein, 49
Left caudal lobe, 68
Left common carotid artery, 51
Left coronary artery, 46, 52, 55
Left coronary vein, 55
Left costocervical artery, 46
Left costocervical vein, 46
Left cranial lobe, 68
Left gastroepiploic vein, 56
Left hemisphere, 84
Left internal thoracic artery, 46
Left lung, 44, 46, 66, 67
Left medial lobe, 68
Left phrenic nerve, 46
Left pulmonary artery, 45, 54
Left pulmonary vein, 46, 54
Left subclavian artery, 46, 48, 51, 52, 54, 55

Left subclavian trunk, 51
Left thoracic cervical trunk, 51
Left vagus nerve, 48
Left ventricle, 44, 45, 46, 49, 50, 51, 52, 54, 55
Lens, 88
Leydig cells, 74, 93
Linea alba, 26
Linguofacial vein, 49, 50
Liver, 35, 36, 37, 38, 41, 56, 57
Lobar bronchi, 68
Longitudinal fissure, 84
Long thoracic vessels, 50
Loop of Henle, 80, 81, 82
Lumbar vertebrae, 10, 11
Lumina, 40
Lung, 35, 36, 44, 46, 66
Lymph nodes, 32, 33, 37

M

Macrophages, 68
Macula densa of distal tubule, 81
Male reproductive system, 72–74
Mammals, 43
Mammary papilla, 6, 7, 15
Mandible, 10, 11, 16, 18, 69
Mandibular gland, 16, 17, 18, 32
Mandibular lymph nodes, 17
Masseter muscle, 16, 17, 18, 33
Maxilla, 10, 11
Maxillary vein, 49, 50
Medial lobe, 66
Medial plane, 2, 3
Medial segment, 66
Median sacral artery, 58, 61
Medula, 80, 81, 93
Medula oblongata, 84, 85, 86
Medullary rays, 81
Menstruation, 92
Mesenteric lymph nodes, 38, 59
Mesenteric vein, 57
Mesentery, 38
Metacarpals, 10, 11
Metatarsals, 10
Middle lobe, 67
Mitral valve, 53
Muscle fiber, 13
Muscle(s)
  adductor, 25, 26, 27, 28
  biceps, 16, 21, 22
  biceps femoris, 18, 19, 22, 24, 28
  brachialis, 16, 19, 20
  brachiocephalicus, 16, 17, 18
  cleidomastoid, 16, 18, 19
  cleidooccipitalis, 16, 18, 19
  coracobrachialis, 22
  cremaster, 73
  deltoid, 18, 19, 20
  diaphragm, 5, 35, 36, 44, 46, 50, 52, 58, 66
  digastric, 16, 17, 18
  extensor carpi radialis, 16, 19, 20, 21
  extensor digitorum communis,

19, 20
  extensor digitorum lateralis, 19, 20
  extensor digitorum longus, 23, 24, 28
  extensor digitorum quarti, 23, 24, 28, 29
  extensor digitorum quinti, 23, 24, 28, 29
  flexor carpi radialis, 16, 20, 21
  flexor carpi ulnaris, 16, 20, 21
  flexor digitorum longus, 23, 25, 26, 28
  flexor digitorum profundus, 16, 19, 20, 21
  flexor digitorum superficialis, 16, 20, 21
  flexor hallicus, 23, 25, 26, 28
  gastrocnemius, 23, 24, 25, 26, 28, 29
  gluteus medius, 18, 19, 23, 24, 28
  gluteus profundus, 27, 28, 29
  gluteus superficialis, 18, 23, 24, 28
  gracilis, 23, 24, 25, 26
  iliacus, 25, 26, 27, 28
  latissimus dorsi, 16, 18, 20
  masseter, 16, 17, 18, 33
  mylohyoid, 16, 17, 18
  oblique, external, 16, 18, 19, 22, 23, 26
  oblique, internal, 18, 22, 23
  omohyoid, 16, 17, 18
  omotransversarius, 18, 20
  pectineus, 25, 26, 27
  pectoralis profundus, 16, 17, 18, 19, 20
  pectoralis superficialis, 16, 17, 20
  peroneus longus, 24, 27, 28, 29
  peroneus tertius, 24, 27, 28, 29
  psoas major, 25, 26, 27
  quadratus femoris, 27, 28, 29
  rectus abdominis, 21, 22
  rectus femoris, 23, 24, 25, 26, 28
  rectus thoracis, 21, 22
  sartorius, 23, 25, 26
  scalenus, 21, 22
  semimembranosus, 18, 23, 24, 25, 26, 28
  semitendinosus, 18, 19, 23, 24, 25, 26, 28
  serratus dorsalis, 18, 22, 23
  serratus ventralis, 18, 20
  soleus, 23, 25, 26, 28
  sternohyoid, 16, 17, 18
  sternomastoid, 16, 17, 18
  sternothyroid, 16, 17, 18
  stylohyoid, 16, 17, 18
  subscapularis, 16, 21, 22
  supraspinatus, 16, 18, 19
  temporalis, 16, 17, 18
  tensor fascia latae, 18, 19, 23, 24, 25, 26, 28
  teres major, 16, 21, 22
  thyrohyoid, 16, 17, 18
  tibialis anterior, 23, 24, 25, 26, 28, 29
  tibialis posterior, 23, 25, 26

  transversus abdominis, 18, 22, 23
  trapezius, 18, 19, 20
  triceps, 16, 18, 19, 20, 21, 22
  ulnaris lateralis, 19, 20
  vastus lateralis, 27, 28, 29
  vastus medialis, 25, 26, 27
Muscular system, 13–29
Muscularis externa, 40
Muscularis mucosae, 40
Mylohyoid, 16, 17, 18

N

Nares, 5, 6, 7, 15, 67
Nasal bone, 10, 11
Nasopharynx, 33, 34, 65, 67, 69
Neck region, 31–33, 43–52, 92
Nephron, 80
Nervous system, 83–89, 91
Nictitating membrane, 87, 88
Normal circulatory pathway, 53

O

Occipital bone, 10, 11
Occipital lobe, 85, 86
Occipital region, 84
Occipital condyles, 11
Oculomotor nerve (III), 85
Olfactory bulb, 85, 86
Olfactory nerve (I), 85
Omohyoid, 16, 17, 18
Omotransversarius, 18, 20
Oocyte, 75
Optic chiasma, 85
Optic disk, 89
Optic nerve (II), 85, 86, 87, 88, 89
Oral cavity, 31–33, 34, 67–69
Origin, muscle, 13
Osteons, 10
Ostium of oviduct, 77
Ovarian artery, 61, 75
Ovarian vessels, 77
Ovary, 75, 77, 92, 93, 94
Oviduct, 75, 76, 77

P

Palate, 33, 34, 69
Pancreas, 36, 38, 39, 40, 41, 56, 57, 59, 92, 93, 94
Pancreatic duct, 38, 39
Pancreatic islet, 40, 93
Pancreatic vein, 56
Papillae, 33, 34
Papillary muscle, 54
Parietal bone, 10, 11
Parietal lobe, 85, 86
Parietal region, 84
Parotid duct, 31, 33
Parotid gland, 31, 33
Pars distalis, 93
Pars intermedia, 93
Pars nervosa, 93
Patella, 10, 11
Pectineus muscle, 25, 26, 27

Pectoral girdle, 10
Pectoralis profundus, 16, 17, 18, 19, 20
Pectoralis superficialis, 16, 17, 20
Pelvic girdle, 10, 11
Pelvic region, 23–29
Pelvis, 61
Penis, 36, 72, 73, 74, 78
Pepsinogen, 35
Pericardial membrane, 43
Pericardium, 44
Perichondrium, 68
Perinuclear sarcoplasm, 14
Peripheral nervous system (PNS), 83
Peritubular arteries, 80, 82
Peritubular capillaries, 81
Peroneus longus, 24, 27, 28, 29
Peroneus tertius, 24, 27, 28, 29
Phalanges, 10, 11
Phrenic nerve, 46
Photoreceptors, 89
Pineal body, 86, 91, 93, 94
Pituitary gland, 85, 86, 91, 93
Pivot joint, 12
Plantigrade, 11
Plicae cirulares, 40
Pons, 85, 86
Portal system, 53
Portal vein, 39, 53, 58
Posterior chamber, 88
Premaxilla, 10, 11
Prepucial gland, 73
Primordial follicles, 93
Primary bronchi, 68
Primary follicles, 93
Progesterone, 92
Proximal convoluted tubules, 80, 82
Proximal region, 2, 3
Proximal tubule, 81
Psoas major, 25, 26, 27
Pulmonary alveoli, 68
Pulmonary arteriole, 68
Pulmonary artery, 44, 45, 46, 47, 48, 52, 55, 66
Pulmonary semilunar valve, 53
Pulmonary trunk, 48, 54
Pulmonary vein, 45, 46, 55, 66
Pulmonary venule, 68
Pupil, 88, 89

Q

Quadratus femoris, 27, 28, 29

R

Radius, 10, 11
Rectum, 40, 41, 58, 74, 77
Rectus abdominis, 21, 22
Rectus femoris, 23, 24, 25, 26, 28
Rectus thoracis, 21, 22
Renal artery, 58, 59, 60, 61, 63, 72, 78, 79, 81, 82
Renal corpuscle, 81

Renal cortex, 74, 81
Renal medula, 81
Renal pelvis, 74, 80, 81
Renal pyramid, 74, 81
Renal vein, 58, 59, 60, 61, 63, 72, 78, 79, 81, 82
Reproductive organs, 5
Reproductive system, 71–78
Reproductive pathway, 6
Reproductive structures, 72
Respiratory epithelium, 68
Respiratory system, 65–69
Respiratory alveoli, 68
Retina, 87, 88
    central artery, 88
Retractor muscle (of penis), 73
Rib cage, 10, 11
Right atrium, 45, 55
Right auricle, 44, 46, 48, 54, 55
Right axillary artery, 51
Right axillary nerve, 51
Right axillary vein, 49
Right brachiocephalic vein, 49, 50
Right caudal lobe, 66, 67, 68
Right common carotid artery, 51
Right cranial lobe, 66, 67, 68
Right gastroepiploic vein, 56
Right hemisphere, 84
Right lung, 44, 46, 66, 67
Right medial lobe, 66, 68
Right phrenic nerve, 46
Right pulmonary artery, 45, 55
Right pulmonary vein, 54
Right subclavian artery, 49, 51, 52
Right subclavian vein, 50
Right thoracocervical artery, 51
Right vagus nerve, 48
Right ventricle, 44, 45, 46, 49, 52, 54, 55
Rostral region, 2, 3
Round ligament, 77
Rugae, 37

S

Sacral vertebrae, 10
Sacrum, 11
Sagittal plane, 2, 3
Salivary glands, 32, 33, 41
Sartorius, 23, 25, 26
Scalenus, 21, 22
Scapula, 10, 11
Sclera, 87, 88
Scrotal sacs, 6, 15, 72
Scrotum, 7, 36, 72, 73, 74
Semilunar valve, 54
Semimembranosus, 18, 23, 24, 25, 26, 28
Seminal vessicles, 72, 73, 74, 78
Seminiferous tubule, 72, 74, 93
Semitendinosus, 18, 19, 23, 24, 25, 26, 28

Sensory neurons, 83
Sensory organs, 5, 87
Seromucous gland, 68
Serratus anterior, 16
Serratus dorsalis, 18, 22, 23
Serratus ventralis, 18, 20
Sheep eye, 87
Shoulder, 6, 20
Skeletal muscle, 13, 14
Skeletal system, 9
Skull, 9, 10, 11
Small intestine, 35, 36, 40, 56, 57, 59, 64
Smooth muscle, 13, 14
Soft palate, 33, 34, 67
Somatostatin, 92
Soleus, 23, 25, 26, 28
Sperm production, 8
Spermatic cord, 72, 73, 74
Spheroidal joint, 12
Spinal cord, 83, 85, 86
Spinal nerves, 83
Spiral accessory nerve (XI), 85
Spiral colon, 35, 38, 40, 41
Spleen, 35, 36, 38, 39, 40, 56, 58, 59, 64
Splenic artery, 64
Splenic nodule, 64
Splenic vein, 56, 64
Sternocephalicus, 19
Sternohyoid, 16, 17, 18
Sternomastoid, 16, 17, 18
Sternothyroid, 16, 17, 18
Stomach, 34, 36, 37, 38, 40, 41, 56, 57, 64
Striated muscle, 13
Stylohyoid, 16, 17, 18
Subclavian artery, 46, 48, 49
Subclavian vein, 49
Sublingual gland, 31, 33
Submaxillary gland, 31, 33
Submucosa, 40
Subscapular vein, 49, 50
Subscapularis, 16, 21, 22
Sulcus, 85
Superficial musculature, 23–24
Supraspinatus, 16, 18, 19
Suture joint, 12

T

Tail, 6, 7
Tapetum lucidum, 89
Tarsals, 10, 11
Teeth, 33, 41
Temporal bone, 10, 11
Temporal lobe, 85, 86
Temporal region, 84
Temporalis muscle, 16, 17, 18
Tendon of tensor fasciae femoris, 28

Tensor fasciae latae, 18, 19, 23, 24, 25, 26, 28
Teres major, 16, 21, 22
Terminal bronchi, 68
Testes, 6, 36, 72, 73, 74, 78, 92, 93, 94
Testicular artery, 61, 63, 72
Testicular vein, 61
Testicular vessels, 74
Testosterone, 92
Thalamus, 86
Third eyelid, 87
Third ventricle, 86
Thoracic arteries, 52
Thoracic cavity, 35, 36, 43–52, 65–67
Thoracic region, 48, 49
Thoracic vein, 46, 49, 50
Thoracic vertebrae, 10, 11
Thoracic vessels, 50
Thoracolumbar fascia, 18
Thorax, 5, 15
Thymosin, 92
Thymus gland, 17, 32, 36, 92, 93, 94
Thyrohyoid, 16, 17, 18
Thyroid gland, 35, 36, 43, 49, 52, 92, 93, 94
Thyrocervical trunk, 52
Thyroxine, 92
Tibia, 10, 11, 25
Tibialis anterior, 23, 24, 25, 26, 28, 29
Tibialis posterior, 23, 25, 26
Tongue, 6, 7, 33, 34, 69
Trachea, 48, 65, 66, 67, 68, 69, 91
Tracheal wall, 68
Transverse plane, 2, 3
Trachea, 65, 66, 67
Transversus abdominis, 18, 22, 23
Trapezius, 18, 19, 20
Triceps, 16, 18, 19, 20, 21, 22
Tricuspid valve, 53, 54, 55
Trigeminal nerve (V), 85
Trochlear nerve (IV), 85
Trunk, 5, 15
Tunical albuginea, 93

U

Ulna, 10, 11, 19
Ulnaris lateralis, 19, 20
Umbilical artery, 15, 35, 36, 47, 58, 60, 61, 62, 72, 74, 75, 76, 77
Umbilical cord, 5, 6, 7, 15, 35, 45, 47, 60, 74, 77
Umbilical vein, 15, 35, 36, 37, 38, 40, 47, 56, 57, 58, 60, 63, 74, 77
Umbilicus, 36

Ureter, 58, 61, 62, 63, 72, 74, 75, 76, 77, 79, 80, 80, 81, 82
Urethra, 6, 72, 74, 76, 77, 78, 79, 80
Urinary bladder, 35, 36, 58, 62, 72, 74, 75, 76, 77, 79, 82
Urinary space, 81
Urogenital opening, 6, 7, 15, 36, 47, 72, 74, 77, 79
Urogenital sinus, 76
Uterine horns, 76
Uterine tube, 77
Uterus, 76

V

Vagina, 76, 77
Vaginal vestibule, 77
Vagus nerve (X), 46, 48, 50, 51, 57, 66, 85
Vastus lateralis, 27, 28, 29
Vastus medialis, 25, 26, 27
Vein(s), 44, 48, 53, 57
    axillary, 21, 49, 50
    azygous, 54, 55
    brachiocephalic, 49
    caudal vena cava, 37, 45, 46, 47, 48, 50, 53, 54, 55, 56, 58, 59, 60, 63, 66, 74, 75, 77
    cranial vena cava, 38, 44, 45, 46, 47, 48, 50, 52, 54, 55, 63
    femoral, 58, 60, 61
    genital, 60
    hepatic, 38, 56
    iliac, external, 58, 60, 61, 63
    iliac, internal, 58, 60
    internal thoracic, 46, 49, 50
    jugular, external, 33, 49, 50
    jugular, internal, 32, 49, 50, 51, 52
    linguofacial, 49, 50
    maxillary, 49, 50
    pulmonary, 45, 46, 55, 66
    renal, 58, 59, 60, 61, 63, 72, 78, 79, 81, 82
    subclavian, 49
    subscapular, 49, 50
Ventral region, 2, 3
Vermis, 85
Vertebral column, 9, 10
Vibrissae, 5, 15
Villi, 40
Visceral muscle, 13
Vitreous chamber, 87, 88
Vitreous humor, 87
Voice box (See Larynx)

W

Wrist, 6, 7

Z

Zygomatic arch, 10, 11